Integrative Statistics
for the Social &
Behavioral Sciences

We dedicate this book to our parents.

Integrative Statistics
for the Social & Behavioral Sciences

Renee R. Ha | James C. Ha

University of Washington

Los Angeles | London | New Delhi
Singapore | Washington DC

Los Angeles | London | New Delhi
Singapore | Washington DC

FOR INFORMATION:

SAGE Publications, Inc.
2455 Teller Road
Thousand Oaks, California 91320
E-mail: order@sagepub.com

SAGE Publications Ltd.
1 Oliver's Yard
55 City Road
London EC1Y 1SP
United Kingdom

SAGE Publications India Pvt. Ltd.
B 1/I 1 Mohan Cooperative Industrial Area
Mathura Road, New Delhi 110 044
India

SAGE Publications Asia-Pacific Pte. Ltd.
33 Pekin Street #02-01
Far East Square
Singapore 048763

Acquisitions Editor: Vicki Knight
Associate Editor: Lauren Habib
Editorial Assistant: Kalie Koscielak
Production Editor: Brittany Bauhaus
Permissions Editor: Adele Hutchinson
Copy Editor: Gillian Dickens
Typesetter: C&M Digitals (P) Ltd.
Proofreader: Theresa Kay
Indexer: Diggs Publication Services, Inc.
Cover Designer: Candice Harman
Marketing Manager: Helen Salmon

Printed in the United States of America

Library of Congress Cataloging-in-Publication Data

Ha, Renee.

Integrative statistics for the social and behavioral sciences / Renee R. Ha, James C. Ha.

p. cm.
Includes bibliographical references and index.

ISBN 978-1-4129-8744-8 (pbk.)

1. Social sciences—Statistical methods.
2. Psychology—Statistical methods. I. Ha, James.
II. Title.

HA29.H145 2012 519.5—dc22 2010051831

This book is printed on acid-free paper.

11 12 13 14 15 10 9 8 7 6 5 4 3 2 1

Brief Contents

Detailed Contents

About the Authors

Renee R. Ha received her Ph.D. in Psychology with a specialization in animal behavior from the University of Washington in 1999. She is currently on the faculty of the Department of Psychology at the University of Washington, where she has taught courses on statistics, introductory psychology, human development, and various animal behavior lecture and laboratory courses. Teaching both undergraduate statistics and subsequent laboratory courses, which required students to apply statistics to data, provided her with important insights into what students need in their introductory statistics course. In 1999, she received the Psychology Department's Distinguished Teaching Award. Her research interests primarily involve wild birds, including social behavior, foraging, and conservation. She travels to Micronesia for her research on endangered birds and studies social behavior of local crows on the beaches of Puget Sound. On personal time, she loves to travel to Europe and to islands in the Florida Keys and the Caribbean.

James C. Ha has a 1989 Ph.D. in Zoology/Animal Behavior from Colorado State University and has been on the faculty of the University of Washington since 1992, where he is a Research Associate Professor in the Psychology Department (Animal Behavior Area). He consults extensively in statistics at the University of Washington and around the world, while his own research examines the social behavior of Old World monkeys and their management in captivity, Pacific Northwest killer whales, local and Pacific island crows, and domestic dogs. He has served as a member of the Animal Behavior Society's Executive Committee in several capacities. He is also certified as a full-level Applied Animal Behaviorist and has his own private practice, Companion Animal Solutions, which deals with behavior problems of dogs, cats, and parrots in the Puget Sound area. He also advises attorneys as an expert legal witness in animal behavior. He travels extensively and speaks on the topics of his research, especially facets of social behavior in both wild and domestic animals, as well as on the treatment of behavior issues in companion animals.

Preface

For many years, we have had both the pleasure and the challenge of teaching undergraduate statistics. We were fortunate to be able to teach both the lecture courses in statistics as well as the laboratory courses where students first begin applying statistics to their own data sets. This gave us a unique view into how well we were getting statistical concepts across. What we learned is that many students were struggling with how to go from the book learning to the practice of choosing the appropriate test, using statistics to interpret the data they collected, and also on how to write up the results for a paper. Since these experiences were universal, even for the best statistics students, we knew that we needed to approach the teaching of statistics in a more user-friendly and applied manner. Fortunately, we had students we were working with in both types of courses, so we knew the specific knowledge gaps to fill in so that students would be more prepared for their laboratory courses. We started inserting more practical materials into the lecture courses and then saw how that changed the preparation of later students coming into their laboratory courses. As the preparation of incoming laboratory students improved, we found that we could expose our laboratory students to even more sophisticated analyses and thus even better prepare them for their future needs for statistics.

We searched widely for textbooks that gave a more practical approach, as well as covering the important concepts, but we were forced to modify each text. Since we were heavily tweaking what sections of the books students should read or skip, we decided to try our hand at writing a book that would meet the needs of our students. Some of the changes we made to more traditional statistics textbooks were the following:

1. We placed correlation and regression toward the end of the book to minimize breaking up the logical flow of material.

2. We exposed students to the general linear model conceptual approach to a degree not currently common.

3. We placed a strong and unique emphasis on "choosing the appropriate test" and on the ability to interpret statistical package output.

4. We presented information about the interpretation and writing of statistical results.

5. We presented information at a medium depth and a high breadth.

6. We chose to increase our emphasis on the conceptual meaning of probability rather than including a great deal of material on the mathematics of probability.

A WALK THROUGH THE ORGANIZATION OF OUR BOOK

So, how did we carry out all of these changes? Well, we used a fairly traditional sequence to present the material, except for correlation and regression. So, the text is organized such that descriptive statistics and basic terminology are in Part I, followed by the groundwork for hypothesis testing in Part II. In Part III, we provide additional coverage of both parametric and nonparametric statistics.

Chapters in Brief Review

Part I

1. Introduction to statistics, our approach, and how to pick a calculator

2. Terms, measurement scales, a debate issue on "Ordinal or Interval?" and your first homework exercises

3. Frequency distributions and graphing details

4. Descriptive statistics, a feature for students on how to use their calculators effectively and on using Excel and SPSS

5. What is a normal distribution, standardized scores such as the SAT, and probability both conceptually and how to calculate discrete probabilities

Part II

1. Sampling distribution of the mean and sampling distributions in general

2. Introduction to inferential statistics and important issues on power and effect size, as well as a debate on whether hypothesis testing is even appropriate

3. Single sample tests, more on power and effect size, degrees of freedom and what they mean, and the introduction of the flowchart

4. Two sample tests, power and effect size, an expanded flowchart, and more step-by-step instructions for Excel and SPSS

Part III

1. Analysis of variance (ANOVA), multiple comparisons, expanded flowchart, options for shorter versus longer hand calculations, and step-by-step instructions for Excel and SPSS

2. More ANOVA designs explained conceptually, with the updated flowchart

3. Correlation and regression, power and effect size, options for shorter versus longer hand calculations, and step-by-step instructions for Excel and SPSS

4. The general linear model, what is it and how it helps you run computer software and interpret output

5. Many nonparametric tests, including when to use them and how useful they are, as well as power and effect size and step-by-step instructions for SPSS

6. Summing up the concepts, review of choosing the appropriate test with the fully expanded flowchart, a real-world example to help you see how this works outside of class, and the modern use of resampling statistics

What is different about our book is in the details of what we include and at what depth and breadth we cover them. We seek to give students the depth and breadth they need to use statistics in the future, and we've tested this book and these methods on more than 1,200 undergraduates at the University of Washington. The students themselves have guided us on what works and what does not work. Here is a preview of some of the highlights that explain where our book differs from other undergraduate statistics texts.

Since in our experience, students traditionally have had a limited exposure to a few statistical tests and primarily learn how to conduct and interpret those tests by hand, we expose students to more tests than they are required to hand calculate, and we introduce output from both Microsoft Excel (beginning in Chapter 2) and SPSS (SPSS Inc.'s Statistical Package for the Social Sciences) beginning in Chapter 4. This allows all students exposure to this material, even if their statistics course does not have access to computers. If computers are available, we also provide step-by-step instructions for how to run the statistics in Excel and SPSS. We chose these software packages because they are much different from one another but commonly used, and SPSS has a student version. We explain the output since there are more numbers provided by computer software packages than you normally use when you hand calculate the same tests. The additional numbers have proven confusing when students get to their laboratory course after they have been trained in the hand calculation techniques. In addition, we include examples of how to interpret and write up the results of each hypothesis test. Unfortunately, the terminology varies among textbooks and software packages. However, the terms make much more sense when one is trained to understand the conceptual basis to the labels, and that is particularly true if you have been exposed to more than one style of output.

Rather than completely throwing out the idea of hand calculations altogether, we emphasize that important calculations be done by hand to help students grasp the underlying concepts on where the numbers come from and what they mean. This can be done for a limited number of tests and still accomplish the conceptual goals of the course.

We introduce the concept of a "general linear model" early in our discussion of inferential statistics. We formally address the relationship between tests and the conceptual basis of a general linear model in Chapter 13, including introducing students to the idea of analyzing data using the general linear model on a statistics package, which is available in SPSS. Without this understanding, students would not be able to analyze data that contain multiple variables that are collected in different ways (categorical data vs. continuous data). We provide the conceptual understanding so that students can apply this more modern technique to appropriate data sets and also better understand the mathematical relationships between tests that are commonly performed in the behavioral sciences (independent and paired t tests from Chapter 9, ANOVA in Chapters 10 and 11, and linear regression in Chapter 12).

Another unique contribution of this textbook is that we present a flowchart that guides students on how to choose the appropriate statistical test for a given data set and experimental design, and that flowchart is built up chapter by chapter as you learn more tests. We include "choose the appropriate test questions" that may encompass statistical tests from previous chapters. In other words, we have some exercises that are cumulative in terms of what test should be used to analyze the data in the story problem. In fact, we feel this is the one of the most important concepts for undergraduate statistics because the computer can do the analysis for you only if you know which statistical test is appropriate. Students commonly come to their laboratory courses without the ability to choose the appropriate test because they have learned statistics one chapter at a time and there was only one statistical test per chapter. When we examine the students in our statistics course at the University of Washington, we make "choosing the appropriate test" a cumulative concept that is fair game on every exam, but we limit the questions on the calculations to the most recent units. We have tried introducing these concepts at earlier and later periods of the term, and our most successful efforts in training students to choose the appropriate test occur when we introduce the flowchart with the first inferential test and when we require that this portion of the course be cumulative. Thus, it is a proven strategy.

We hope that you will find our book to be approachable but rigorous in its ability to show you how statistics are chosen, carried out, interpreted, and written up in a manuscript. We have worked hard to check and double-check for errors and typos; if some slipped through, please feel free to e-mail us at jcha@uw.edu or robinet@uw.edu. Also, we would welcome hearing about your experiences in using the book.

Acknowledgments

Our undergraduate students at the University of Washington inspired us to write this textbook and have taught us a great deal about how to teach statistics as clearly as we possibly can. The authors would like to thank our many past undergraduate and graduate teaching assistants who have contributed ideas, calculations, and inspiration to the homework problem sets. We would also like to thank the many anonymous reviewers who made many helpful suggestions . . . some of which we incorporated! Thanks especially to Carissa Leeson, who read several versions of this textbook and helped R. Ha clean up her grammar and flow. Any remaining errors are clearly ours and not theirs! We would also like to acknowledge the Whiteley Center at the University of Washington Friday Harbor Laboratories, which provided facilities and a quiet place to write. We wrote the first few chapters on another island, in the Florida Keys, and we thank everyone who made that possible. Andrew Ha provided the suggestion for the example about mushy Valentine's Day cards many years ago, when girls were dumb. Thanks to Peter and Angie Ha, who took the photo of us in Seattle. R. Ha would like to thank her mother, Brenda, for encouraging her to pursue her education, and her father, Steve, for supporting her writing aspirations. While he passed away during the final stages of publication, we know he would have been thrilled to see this in print. J. Ha would like to thank his parents, Sam and Margaret, for always supporting his academic pursuits. Finally, when we were at our lowest point in the efforts to publish our work, a wonderful editor named Vicki Knight appeared to save us and finally get our work to print. Thanks also to all of the staff at SAGE for making this process so smooth and professional.

The authors and SAGE gratefully acknowledge the contributions of the following reviewers (and their anonymous reviewers):

Janis E. Johnston, *American Association for the Advancement of Science, Science and Technology Policy Fellow*

Komanduri S. Murty, *Clark Atlanta University, Fort Valley State University*

Benjamin Pearson-Nelson, *Indiana University, Purdue University Fort Wayne*

Iris Phillips, *University of Southern Indiana*

Carter Rakovski, *California State University, Fullerton*

Thomas Sanocki, *University of South Florida*

PART I

DESCRIPTIVE STATISTICS

CHAPTER 1
Introduction to Statistics
What Are You Getting Into?

WHAT IS STATISTICS?

This book addresses the topic of statistics, but what exactly do we mean by statistics? Statistics is a branch of applied mathematics that involves the collection and interpretation of data and the use of mathematical principles to draw conclusions about the results of our observations. Stated another way, statistics is a field of mathematical study that addresses how data that are collected in a study (such as the number of accidents at a particular intersection) can reveal patterns and how to evaluate those patterns based on the likelihood of that pattern occurring by chance. Patterns that are considered to be unlikely occurrences if chance alone is operating are then said to be occurring for reasons other than chance (such as poor visibility in the intersection or some other variable that we believe might be affecting the outcome of our study).

WHAT IS OUR APPROACH TO STATISTICS?

This book is intended for those of us who use statistics as a tool to understand the data that they collect in their research or to understand the data collected by others in their research. Thus, we will not delve too deeply into the theoretical basis behind the mathematics that we teach, but you will definitely be using some of your skills in mathematics in this course, and it would be a good idea for you to review some basic college mathematics textbooks if you feel you have difficulty recalling things like the proper order that you follow when you have multiple operations in a single problem (e.g., always do what is in the parentheses first). There will be statistical tests that you will learn to conduct by hand. However, while we feel that requiring some calculations to be done by hand is important in teaching the underlying basis of statistical interpretation, this can be done for a limited number of tests and still accomplish the conceptual goals of the course.

In addition, we will emphasize how the steps you take in calculating and evaluating a statistical test by hand compare to interpreting a printout of the same statistical test in two different computer software packages. Because the statistical terminology varies among textbooks and software packages, it is critical that you are trained to understand the conceptual basis to the labels. That will be easier for you if you have been exposed to more than one style of output. Truthfully, very few scientists calculate statistical tests by hand, and when they do, it is only for the simple tests. Thus, we will provide examples in the appropriate chapters that demonstrate the step-by-step calculations done by hand and then show the output from the same data in Microsoft's Excel and SPSS Inc.'s Statistical Package for the Social Sciences (SPSS). We chose these software packages because they are much different from one another but commonly used, and SPSS has a student version. We will highlight the numbers that match the numbers that we calculated by hand, and we will define all of the output since there are more numbers provided by computer software packages than you normally use when you hand calculate the same tests. The additional numbers have proven confusing to students when they get to their laboratory course if they have only been trained in the hand calculation techniques.

Beginning in Part II, you will be introduced to a flowchart that will guide you on how to choose the appropriate statistical test for a given data set and experimental design, and that flowchart will be built up chapter by chapter as you learn more tests. In fact, we feel that choosing the appropriate test is the most important concept in undergraduate statistics because the computer can do the analysis for you only if you know which statistical test is appropriate. Students commonly come to their laboratory courses without the ability to choose the appropriate test because they have learned statistics one chapter at a time and there was only one statistical test per chapter.

We will also introduce you to the concept of a "general linear model." This is often neglected in an undergraduate course, but it is an important and useful concept. On the basis of our coverage of this model, you will be able to analyze data using the general linear model on a statistics software package (such as SPSS) if you ever need to do so. You will be able to apply this more modern technique to appropriate data sets and will also better understand the mathematical relationships between tests that are commonly performed in the behavioral sciences (independent t test, paired t test, analysis of variance [ANOVA], linear regression), all of which you will hear about in later chapters.

PRACTICAL DETAILS BEFORE WE CAN MOVE ON

Before we move on into new material in Chapter 2, there are a couple of practical details to cover, including a discussion of rounding and features of a statistical calculator. The details of this material might be altered by your instructor for the specific details of your class, but this information should prove useful for most students.

A Short Discussion of Rounding

The rules about rounding vary a bit based on the field of study. We will follow the rule that is common in the social sciences, which is to round to two decimal places. For example, if you have a number such as 164.0665, we would round that number using rules that are based on the number in the third decimal position.

Rounding Rules

If the number in the third decimal position is > 5, we would round the second decimal position up by one unit. If the number in the third decimal position is < 5, we would retain the number in the second decimal. If the number was exactly 5, then we would look at the next decimal place (fourth) but apply the same "less than or greater than 5" rule.

Example

For the number 164.0665, the number in the third decimal position is a 6, and thus we would round the number in the second decimal position (also a 6) up one unit (7). Thus, our newly rounded number would be reported as 164.07.

However, the number 164.0652 would be reported as 164.06 since you are forced to go to the fourth decimal number and it is smaller than 5. This technique may differ from the one that you were taught earlier in your education, but following this technique helps avoid bias in rounding that can occur if you round up simply based on the number in the third position being equal to 5.

A Warning About Rounding

Keep in mind that you should not round numbers until you arrive at your final calculation. Rounding as you calculate intermediate numbers can result in the accumulation of small errors.

Buying an Appropriate Calculator

You should avoid buying a calculator that will do more than you need to do in this course. Students often buy "programmable" or "graphing" scientific calculators, and they not only are unnecessary for this course but also often confuse the student who is trying to use them. There are a few symbols that you should look for in a calculator. The symbols sometimes vary by the manufacturer, but we will give you some of the more common variations of the symbols you need to aid you in your purchase. The calculators that we (as instructors) use for this course cost us between U.S.$6 and $8 and were purchased in an office supply store, so you can save yourself some money as well as frustration by following our advice. You do need to find a scientific or statistical calculator and not a business calculator, but you will likely want to stay away from the programmable calculators unless you are extremely proficient with one already.

Symbols You Need on Your Calculator

Σ or a Memory Storage Key, such as M+ or Σ+

ΣX

ΣX^2

$\bar{X}$ or μ

s_n or σ or σ_n

s_{n-1} or s or σ_{n-1}

$\sqrt{}$

X^2

How to Use Your Calculator

It is difficult to give you instructions on how to use your calculator since they vary so much, but there are excellent resources on the Internet to help you with this task. It is very important to take the time to learn how to enter data and to use the memory storage function on your calculator. This will save you significant time and errors on your homework and exams.

SUMMARY

This book will focus on the conceptual and practical aspects of statistics, particularly on how they are used in the behavioral sciences. You should heed our advice to purchase an inexpensive calculator that is easy to use and will aid you in performing well in this course. In the next chapter, we will introduce you to the important terminology used in statistics.

CHAPTER 2

Getting Started With Statistics

Basic Terminology

Descriptive Versus Inferential Statistics

Statistical Notation

Measurement Scales

Summary

Chapter 2 Homework

It's common for students to approach their introductory statistics course with some trepidation, but we believe that you will find this material easier than you expected, providing that you prepare yourself along the way. One of the best ways to do this at the beginning of the course is to thoroughly understand the basic terminology. What may appear foreign to you initially will become second nature to you, providing you make yourself comfortable with the terms that are unique to this field. Our job is to make this process as easy as we possibly can.

In this chapter, we provide definitions of statistical terms, familiarize you with statistical notation, and provide some examples of common statistics. In addition, you'll find some output from Microsoft Excel that demonstrates some basic statistics you can produce using this software. We use Microsoft Excel to demonstrate computer output because Microsoft Excel is readily available on college campuses and on most home computers.

BASIC TERMINOLOGY

Behavioral scientists are interested in questions such as the relationship between life stressors and domestic abuse or the factors that influence reading scores in grade-schoolers. In a perfect world, we would measure every instance of domestic abuse and the reading scores of every grade-schooler, but obviously that is rarely possible. Thus, while we would like to work with an entire **population** to know the complete set of individuals, objects, or scores of interest, it is much more common that we are forced to work with a subset of our population of interest. This subset is termed our **sample.** In fact, there is no reason to require us to determine the population of scores: Statistics allow us to draw conclusions on the basis of much more feasible samples instead.

> **Population:** The complete set of individuals, objects, or scores that is under study.

> **Sample:** A subset of the population of interest.

For example, Smith and Smoll (1990) studied the effect of coaching style on self-esteem in children. In general, they were interested in the impact that various coaching styles had on children's self-esteem, but measuring every coach's behavior and the resulting impact on every child in youth sports would clearly be impossible. Thus, while their population is all youths that participate in sports, their sample was actually 542 male athletes participating in a Little League program in the Seattle area (see Figure 2.1).

Figure 2.1

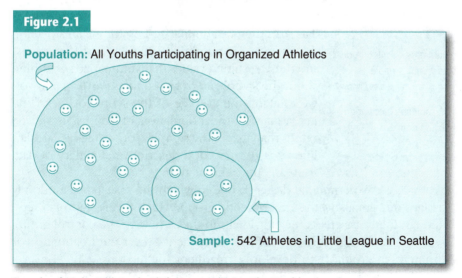

Population: All Youths Participating in Organized Athletics

Sample: 542 Athletes in Little League in Seattle

Sample of Male Athletes in Little League in Seattle, Washington, From a Larger Population of All Youths in Athletic Programs

To study psychology, scientists must measure their subjects in some quantitative or qualitative fashion. Thus, they measure properties of behavior, physiology, socioeconomic status, and so on, which may take on different values at different times and change with various conditions. Some of the **variables** in the Smith and Smoll (1990) study included measures of a coach's instructiveness and supportiveness as well as the level of self-esteem in the athlete. In contrast to a variable, a **constant** is a value that does not change under different conditions: In our discussions of statistics, variables will generally appear as fixed values in some formulas. **Discrete variables** do not have units in between them, such as hair color (blonde, brunette, red) or political party affiliation (Democrat, Republican, Libertarian). **Continuous variables** have an infinite number of possible values between units, such as temperature (30 degrees, 31 degrees, 30.5 degrees, 30.54672 degrees) or concentrations of a hormone in the blood (2.343 picograms per milliliter). More difficult to fit into such clear categories are variables that are integer measures, such as the number of errors in navigating a maze (one error, two errors, but someone cannot make 2.3 errors): These are generally treated as if they were continuous variables.

> **Variable:** Any property that may take on different values at different times and may change with various conditions.

> **Constant:** A value that is fixed.

> **Discrete variable:** A variable that has no values in between each unit.

> **Continuous variable:** A variable that can have an infinite number of values between each unit.

As the measurements are collected, the resulting values are commonly called **data.** These values, or data, can then be summarized by, for instance, taking the average or examining the high and low values, and these resulting values that summarize something about the sample are referred to as **statistics.** If you were taking the average of the scores of *all* of the individuals in the population of youths in sports, rather than just of the sample, you would be calculating a **parameter,** or a descriptive summary of data from an entire population.

> **Data:** A measurement of a variable.

> **Statistic:** A summary calculation on data collected from a sample.

> **Parameter:** A summary calculation on data collected from a population.

A classic experimental design is one in which the researcher randomly assigns individuals to the experimental group, which receives a treatment, and randomly assigns other individuals in the sample to the control group, which experiences all aspects of the experiment *except* the treatment. For example, in an effort to develop a treatment for nicotine addiction in a sample of smokers, you might assign some participants to an experimental group that views a video on the negative psychological and physiological effects of smoking and assign other participants to a control group that views a video on an eye surgery (a topic presumably having nothing to do with smoking but a video viewing, to keep conditions as similar as possible).

In this experimental design, the property that the researcher either manipulates experimentally or records as a naturally changing condition is described as the **independent variable.** A clear distinction is that the independent variable is the predictor variable. The result of the experiment that the researcher is interested in measuring is defined as the **dependent variable.** The dependent variable is often referred to as the outcome variable.

> **Independent variable**: A predictor property that the researcher either manipulates experimentally or records as a naturally changing condition (thus, a quasi-independent variable).

> **Dependent variable**: An outcome property that the researcher measures.

In our example, the independent variable is the group (control vs. treatment). Another type of independent variable might be the number of relatives of a smoker who have suffered from smoking-related illnesses. In this case, the researcher would be recording a variable that varies "naturally," or a quasi-experimental independent or predictor variable. Your dependent variable, in either case, might be the number of cigarettes smoked by the individuals in each group in the 7 days following the viewing of the respective video or in the 7 days following the interview to obtain "naturally varying" independent variables.

Note that in the example on tobacco use, when the researcher is creating the experimental treatments (watching the videos), he or she must randomly assign participants to each group or condition. Random assignment is defined as assignment of subjects or participants that ensures that subjects are equally likely to be assigned to the treatment or the control condition and that each group of subjects accurately reflects the larger population of subjects. This makes it less likely that the two groups will differ significantly from one another prior to watching the video. It would be a poor design if you were to allow participants to assign themselves to each condition on their own will. Clearly, participants who were more motivated to quit smoking prior to the study would elect to be in the treatment condition, while those who were not ready or willing to quit smoking would sign up for the control condition. This alone could mean that the two groups show a difference in the average number of cigarettes smoked in the week following the videotape, and the purpose of the experiment is to determine whether the treatment alone is effective in reducing smoking in a random sample of individuals taken from the entire smoking population.

In theory, it is not only important to randomly assign participants to each condition but also necessary to randomly select participants from the population of smoking individuals. A failure to randomly select from the population as a whole could result in findings that cannot be generalized back to the entire population because the individuals are not typical of the population. It is rare for scientists to study a sample for its own sake: The goal of observing a sample while being able to generalize findings back to the population is a common one. Having said this, it is logistically nearly always impossible to actually randomly select from the (usually very broad) population that psychologists would like to incorporate into their conclusions. This is an important issue in some studies and one that needs to be considered in interpreting the results of any research study: Does the sample of participants considered in the study accurately reflect the important characteristics of the population to which the conclusions are being applied?

DESCRIPTIVE VERSUS INFERENTIAL STATISTICS

> **Descriptive statistics:** Numbers that summarize a set of data in one of four ways, including a measure of central tendency, a measure of the variability of the scores, a measure of the shape of the distribution of scores, and a measure of the size of the sample.

This book is organized so that you learn about descriptive statistics prior to learning about inferential statistics. **Descriptive statistics** are numbers that summarize your data; in other words, they are numbers that help describe your data set. These include the average (or mean) and the range (largest score minus smallest score), which you have most likely encountered in earlier mathematics, as well as the standard deviation and the shape of the distribution of scores, which you may not have encountered previously. We will save our discussion of standard deviation and the distribution of scores for another chapter, but it is important to understand that all of these descriptive statistics attempt to describe a larger group of scores (your data) with a single number or category, such as an average, or visually by graphing the data as a frequency distribution.

For example, the following numbers are difficult to summarize verbally but can easily be summarized with descriptive statistics:

65
6
23
11
15
241
9
55
1

2
26
2
48
8
41
36
81
9

The average, or **mean,** represents the central tendency of the data (more on this in Chapter 4) and is equal to the sum of the scores divided by the number of scores. In this case, the sum is 679 and the number of scores is 18, so $679 \div 18 = 37.72222$ (see Table 2.1). The smallest score is 1, and the largest score is 241. If you subtract the smallest score from the largest score, $241 - 1 = 240$, you get the **range** of scores. The mean is an example of a statistic that summarizes information about the "central tendency" of the data, while the range summarizes an aspect of the "variability" of the data around the center. The count ("*N*") is another descriptive statistic. Other statistics can summarize information about the "shape" of the scores or the pattern of the scores around the center of the distribution of scores. We will present these statistics in Chapter 4.

Inferential statistics are numbers that we calculate based on sample data but that we are attempting

> **Mean:** A statistic that measures the middle, or central tendency, of a set of scores and is calculated by taking the sum of the scores and dividing by the number of scores.

> **Range:** A statistic that measures the variability of a set of scores and is calculated by taking the largest score and subtracting the smallest score.

> **Inferential statistics:** Use of sample statistics in an attempt to infer the characteristics of a population of scores and ultimately to make decisions about experimental hypotheses.

Table 2.1	
Mean	37.72222
Range	240
Minimum	1
Maximum	241
Sum	679
Count	18

Modified Excel Summary Statistics Table

to use to generalize to a larger population of data that we have not sampled. These statistics require us not only to calculate summary statistics on sample data but also to estimate population values that are unknown or unavailable and then to evaluate the likelihood that the sample represents the population. The first step in learning how to perform inferential statistics is to learn about descriptive statistics, and thus we will save our discussion of inferential statistics for Chapters 6 and 7.

STATISTICAL NOTATION

Do not make the mistake that many of our students have made in this course. These students have underestimated the importance of learning the statistical notation and, as a result, have later felt that we were speaking to them in Greek (which, symbolically, we were!). Unfortunately, this meant that they struggled to understand the material in this book more than they should have. Learning the terminology and the notation in this chapter will give you the vocabulary to understand the rest of the course.

We typically represent the variable that the scientist is measuring with an X or a Y and the number of measurements the scientist has taken (or the size of the sample) by N or, later in the text, by lowercase n. Capital N represents the number of scores in a population, while lowercase n is the number of scores in a sample, or the "sample size." Subscripting is used to identify individual scores, such as X_1 (first value of X in the list), X_2 (second value of X in the list), X_3, and so on all the way to the last score, which can be symbolized as X_N. Now you have the tools you need to understand how to sum all of the scores. This technique is termed *summation* and essentially means to add up all of the scores or at least to add up the scores that are indicated by the notation.

The symbol for summation is $\sum$ (uppercase Greek letter sigma), and it indicates that you should add up the scores. If you wished to add up *all* of the scores in the sample, you would indicate that in notation in the following way:

$$\sum_{i=1}^{N} X_i.$$

This notation tells one to add up all of the scores from the first score (let $i = 1$) to the last score (when $i = N$) for each individual score (X_i).

In practice, these elaborate notations are often dropped so that the notation simplifies to

$$\sum X.$$

This literally says that you should add up all of the X scores. For example,

$$\sum X = X_1 + X_2 + X_3 + \ldots X_N.$$

If we use the sample data that we used to demonstrate descriptive statistics earlier in the chapter, we would calculate the sum of those scores in the following way:

$$\sum X = 65 + 6 + 23 + \ldots 9 = 679.$$

Note that we have not written out every score in this equation, but you would need to add them all together, and if you do so, you will find that you get the same sum that we did when we used Excel to calculate the sum rather than calculating it manually. Refer to Table 2.2 for a list of all of the scores.

Table 2.2

Symbolic Score	Actual Score Data	Squared Scores
X_1	65	4,225
X_2	6	36
X_3	23	529
X_4	11	121
X_5	15	225
X_6	241	58,081
X_7	9	81
X_8	55	3,025
X_9	1	1
X_{10}	2	4
X_{11}	26	676
X_{12}	2	4
X_{13}	48	2,304
X_{14}	8	64
X_{15}	41	1,681
X_{16}	36	1,296
X_{17}	81	6,561
X_{18}	9	81
	$\sum\limits_{i=1}^{N} X_i = 679$	$\sum\limits_{i=1}^{N} X_i^2 = 78,995$

Sample Data Shown Symbolically and Numerically

Another common use of the summation symbol in statistics is to show that the scores should be squared and then added to one another. This is most often represented as

$$\sum X^2 \, ,$$

but of course it should actually be written out in the following way to be complete:

$$\sum_{i=1}^{N} X_i^2 \, .$$

When this command is elaborated with the scores used in the previous example, you get the following:

Symbolically

$$\sum_{i=1}^{N} X_i^2 = X_1^2 + X_2^2 + X_3^2 + \ldots X_N^2 \, .$$

With the Sample Data

$$\sum_{i=1}^{N} X_i^2 = (65)^2 + (6)^2 + (23)^2 + \ldots (9)^2 = 78,995 \, .$$

With the Squared Values of the Sample Data

$$\sum_{i=1}^{N} X_i^2 = 4,225 + 36 + 529 + \ldots 81 = 78,995 \, .$$

This is called the "sum of the squared Xs." This concept can be confused with the "sum of the Xs squared." The second case refers to adding up all of the scores and then squaring the sum of those scores.

Symbolically

$$\left(\sum_{i=1}^{N} X_i \right)^2 = (X_1 + X_2 + X_3 + \ldots X_N)^2 \, .$$

With the Sample Data

$$\left(\sum_{i=1}^{N} X_i \right)^2 = 65 + 6 + 23 + \ldots 9 = (679)^2 = 461,041 \, .$$

Notice that taking the sum of the Xs squared results in a much larger number (461,041) than you get when you take the sum of the squared Xs (78,995). The wording of these two sums is quite confusing, but you can dispense with that confusion by carefully noting where the squaring takes place. If the notation is simplified, as it often is in practical use, then it is a matter of differentiating between $\sum X^2$ and $(\sum X)^2$. The first symbol tells you to sum the squared scores, and the second symbol tells you to square the sum of the scores! You can use the exercises at the end of the chapter to practice reading these different symbols and carrying out the appropriate operations.

MEASUREMENT SCALES

Scientists collect different kinds of data. In this case, we don't necessarily mean different variables: Obviously, they do that. Rather, there are different scales or types of data, and the scale on which data are recorded can significantly affect the hypotheses that can be tested and how to test those hypotheses. As an example of similar dependent variables but different scales, you might be collecting data on the number of days in a month that a depressed individual felt overwhelmed or hopeless about his or her life. A similar measurement of depression would be to ask individuals in your sample to rate themselves on a scale from 1 to 5 on how depressed they felt on average, with 1 being not depressed and 5 being very depressed. These sound like very similar measurements, and they are both attempts to measure underlying depression, but they are actually using two different measurement scales to collect the data for later statistical analyses. It turns out that there are not two but four measurement scales that are used to collect data (dependent variable) and to design experiments (independent variable), and which ones you use determine the mathematical manipulations and the statistical tests that you can use on the data later. Not only is this distinction critical to understanding what inferential statistical tests you can perform later, but it is also a concept that appears deceptively simple initially.

These measurement scales vary in their possession of three properties: magnitude, equal intervals, and absolute zero. *Magnitude* refers to having a quantified measure of the amount of something: A measure with the property of magnitude actually reflects a physical amount of the property in question. The term *equal intervals* refers to a scale in which the amount of the property being measured is the same at all places in the scale. For example, the difference in the amount of the hormone estrogen from 2.3 milligrams to 2.4 milligrams (0.1 milligrams) is the same amount of estrogen as the difference between 10.8 milligrams and 10.9 milligrams. This cannot be said of, say, a rating system for football teams: first place, second place, and third place. The difference in ability between first- and second-place teams is not necessarily the same as the difference in ability between the second- and third-place teams: That's what "a distant third" refers to. *Absolute zero* refers to the property in which "zero" on a scale actually refers to the physical absence of something: zero degrees Fahrenheit does not reflect the absence of heat; it is an arbitrary zero point. Zero errors through a maze does reflect the absolute absence of errors, and therefore maze errors would be described as a scale having an absolute zero.

Examples of Each Measurement Scale

> **Four Measurement Scales**
>
> Nominal: Data are on a categorical, and often qualitative, scale rather than one that is quantitative.

Nominal data are collected whenever you ask how many students in a class belong to each political party (Democrat, Green, Independent, Republican, or Other), and you simply count the number (or frequency) of students that fall into each group. Your data might look something like Table 2.3.

In this example, students are placed into categories of political parties. These categories cannot be meaningfully ranked, and they do not include information on magnitude (greater or less than) or a true zero point. The term *nominal* means "name," and nominal data are often referred to as "naming scales." Other examples of data collected on a nominal scale include the number (or frequency) of individuals in a sample who are wearing various types of running shoes (Adidas, Nike, Reebok, etc.) or the number of individuals in a sample or population who belong to various ethnic groups (African, Asian, Caucasian, Middle Eastern, etc.) or have different blood types (A, B, AB, O). Again, these are qualitative categories that cannot be ranked in order and do not include measures of magnitude.

> Ordinal: Data are on a categorical scale, in which categories can be ranked in relative order.

Ordinal data do include information on rank or order. Some examples of this scale include the order that runners finish in a race (first, second, third, fourth, fifth, etc.). While this scale obviously tells you who performed the best in this case, it does not contain any

Table 2.3				
Democrat	**Green**	**Independent**	**Republican**	**Other**
31	4	8	29	1

Sample Data for the Nominal Scale

information on the magnitude of the differences between the ranks. You do not know whether first and second place was decided by 10 minutes or by 2/100th of a second. In the second scenario of our earlier depression example, the researchers asked participants to rank their average level of depression on a scale from 1 to 5. Data collected in this manner is on an ordinal scale and is common in the behavioral sciences, whether you are studying people or animals. An animal behaviorist might record the levels of aggression in a baboon troop by using a similar scale.

Data collected on an **interval** scale have information on order or rank like the ordinal scale, but the data are more quantitative because they contain information on magnitude. Thus, the scientist knows not only that one score is higher or lower than another but exactly how much higher or lower one

> **Interval:** Data are collected in a manner that measures actual magnitude and has equal intervals between possible scores but does not have a meaningful absolute zero point.

score is from another score. Marine biologists recording daily water temperatures (in Celsius) on the Great Barrier Reef are using an interval scale. Both Celsius and Fahrenheit temperature scales include information on magnitude: 23°C is warmer than 20°C, and it is warmer by 3°C. Another property of interval scales is that they have equal intervals between measurement points. In other words, the magnitude of difference between 1°C and 2°C is the same amount of heat (or energy) as the difference between 30°C and 31°C. However, while you can measure temperatures of 0°C or 0°F, these measures are relative zero points rather than absolute zero points. What does this mean? Absolute zero is the absence of temperature or, more accurately, the absence of energy. It turns out that although 0°C is the temperature at which water freezes, it is not the absence of all heat or energy. Likewise, 0°F does not measure the absence of energy. Thus, both of these temperature scales measure on an interval scale, but not a ratio scale. To be a **ratio** scale, they would have to contain a meaningful absolute zero point. You can measure temperature on a ratio scale, the Kelvin scale. In this case, 0° Kelvin is the absence of

> **Ratio:** Data are collected in a manner that measures magnitude, has equal intervals between possible scores, and contains an absolute zero point.

heat or energy. Thus, it contains a meaningful zero point as well as the properties of magnitude and equal interval. Distance can also be measured on a ratio scale (inches, centimeters, miles, kilometers, etc.). Distance is ratio rather than interval because you can have, at least theoretically, zero distance between objects. In the first scenario of the depression example we gave earlier, where participants were asked to tell the researcher how many days a month they felt depressed, we were collecting data on a ratio scale. In this case, reporting more days per month means that a participant feels more depressed, and these data have equal intervals (1 day) and an absolute zero point (participants who are never depressed report zero days feeling depressed). In practical terms, this is a better measure of depression than the earlier ranked (1–5) scale because it contains more quantitative information that allows a statistician more choices in later statistical analyses.

Overview of Measurement Scales

We have summarized the properties of each measurement scale in Table 2.4. Some of our students have found this method of presenting the information more useful than the definitions, and many students find this table and the examples to be invaluable in understanding the practical differences between the measurements scales. Fortunately, in practice, statisticians do not discriminate between interval and ratio scales since they both contain the necessary properties for all types of inferential statistics.

Table 2.4

Scale	Order	Magnitude	Equal Intervals	Absolute Zero	Discrete or Continuous
Nominal	No	No	No	No	Discrete
Ordinal	Yes	Some	No	No	Discrete
Interval	Yes	Yes	Yes	No	Continuous
Ratio	Yes	Yes	Yes	Yes	Continuous

Summary of the Properties of the Four Measurement Scales

SUMMARY

In this chapter, we have presented the basic terminology that is needed to understand the rest of the material in this textbook. We can't emphasize enough the value of learning this material very well, right now. To repeat what we said earlier, for instance, one of the seemingly easiest concepts to (apparently) learn, yet most difficult to apply, are the measurement scales. The concept of different measurement scales will become very important later in this textbook, and you will need to apply the concept to real experimental situations (the dreaded word problems!). So start learning these concepts early and learn them well. The payoff later will be significant.

ISSUES FOR DEBATE: ORDINAL OR INTERVAL?

There are some measures for which there is controversy about the scale of the measurement. For example, the Likert scale refers to a widely used questionnaire format named for its developer. In the questionnaire, participants are asked to choose from several responses in a range such as *strongly agree, agree, undecided, disagree,* and *strongly disagree.* Each response is then associated with a number (e.g., 1–5).

Traditionally, this scale has almost always been treated as if it represents an underlying interval property. However, a growing body of statisticians argues that the interval between these categories is not necessarily equal. In other words, the difference between *strongly agree* and *agree* may not be the same level of "agreeness" as the difference between *agree* and *undecided* and assuming that they are equal may break the assumptions of some inferential statistical tests.

Evidence that treating Likert scales as ordinal can lead to erroneous statistical conclusions is beginning to appear for some measures and statistical tests (see Nanna & Sawilowsky, 1998, for a further discussion of this issue in statistics). Nevertheless, this issue continues to be a controversial one, and many respected scientists advocate treating the Likert scale as an interval scale.

CHAPTER 2 HOMEWORK

Explain the difference between the following terms.

1. What is the difference between a population and a sample?

2. What is the difference between a variable and a constant?

3. What is the difference between an independent variable and a dependent variable?

4. What is the difference between a statistic and a parameter?

5. What is the difference between continuous variables and discrete variables?

6. What is the difference between descriptive and inferential statistics?

Provide an explanation for the following questions.

7. Why is random sampling so important?

8. Why is it important to discriminate between measurement scales?

Identify the following terms in the story problems:

- The independent variable
- The population
- The dependent variable
- The sample
- The statistic

9. A group of researchers want to test their hypothesis that sleep deprivation will decrease the number of lectures attended in a quarter. The researchers round up 350 students from the University of Washington's 38,000 undergraduates and randomly assign each student to either the 4-hours-of-sleep-per-night group or the 8-hours-of-sleep-per-night group. Each participant's attendance is recorded for one quarter, and the mean attendance days per group are calculated.

10. One hundred fifty high school students from Santa Monica are tested to see if teacher gender affects test scores for California high schoolers. Half of the students learn material for a test from a male teacher, and the other half learn the same material from a female teacher of equal teaching ability. Mean test scores are compared.

Define the following terms.

11. What is the nominal scale?

12. What is the ordinal scale?

13. What is the interval scale?

14. What is the ratio scale?

15. Macy's keeps a detailed list of the different brands of shoes its sells (e.g., Doc Marten's, Nike, Nine West). What type of data is this?

   A. nominal
   B. ordinal
   C. interval
   D. ratio

16. The Kelvin scale is a(n) _____ scale, while the Fahrenheit scale is a(n) _____ scale.

   A. nominal, ratio
   B. interval, ordinal
   C. ratio, interval
   D. interval, ratio

17. Which of the following scales correctly describes its associated information?

   A. nominal: time it takes to finish these homework questions
   B. ordinal: a 1 to 7 ranking about the helpfulness of these questions
   C. interval: the names of different caffeine-filled drinks you might consume while trying to stay awake through these questions
   D. ratio: confidence level about the first test

Using the following data points, calculate the answers to the questions below:

$$x_1: 72; x_2: 29; x_3: 10; x_4: 49; x_5: 22; x_6: 58; x_7: 63$$

18a. $N = ?$

18b. $\sum X = ?$

18c. $\sum X^2 = ?$

18d. $\left(\sum X\right)^2 = ?$

18e. $\sum(X - 6) = ?$

18f. $\sum X + 12 = ?$

For Questions 19 to 23, identify the scale (nominal, ordinal, interval, or ratio) for each of the following variables:

19. The time it takes to react to the sound of your alarm clock in the morning.

20. Anxiety over public speaking scored on a scale of 0 to 100. (Assume the difference in anxiety between adjacent units throughout is not the same.)

21. Proficiency in mathematics scored on a scale of 0 to 100. (The scale is well standardized and can be thought of as having equal intervals between adjacent units.)

22. The temperature of a patient measured in degrees Fahrenheit.

23. The birth weight of a premature infant as measured in pounds and ounces.

24. Sex of children is an example of a(n) _____ scale.

    A. ratio

    B. nominal

    C. ordinal

    D. interval

25. Which of the following variables has been labeled with an incorrect measuring scale?

    A. the number of students in a public health class—ratio

    B. ranking in a beauty contest—ordinal

    C. finishing order in a poetry contest—ordinal

    D. self-rating of anxiety level by students in a statistics class—ratio

CHAPTER 3

Frequency Distributions and Graphing

In this chapter, we introduce frequency distributions and concepts related to looking at data graphically. This concept has significant practical implications for all of us. We are presented data in graphical form in newspapers and popular news magazines, as well as on television news broadcasts. Many of us know that these graphical representations can "lie" with the use of truncated axes that make small differences appear very large. This is one area of graphing that it is good to know about, but you can also use graphs (properly done) to aid in validly interpreting complex statistical data.

We will begin the chapter with some typical ways to organize or group data into categories in tabular format, and then we will turn to graphing. The graphs in this chapter are used to provide a summary of data, and many of them are incorporated into determining whether certain assumptions of inferential tests have been met. For example, the section near the end of this chapter on the shapes of frequency distributions is critical to the use of inferential statistics. This material is laying the groundwork for the more sophisticated statistical analyses that you'll be doing later in the course.

FREQUENCY DISTRIBUTIONS IN TABLE FORMAT

Frequency distribution: Data are collapsed into classes or ranges of values, and then the number of observations (or frequency) of each class, or range of values, is recorded in a table or graph.

Frequency (f): The number of scores in that class interval.

Frequency distributions are a useful way to summarize raw data. Typically, it is difficult to see patterns in pages and pages of raw data or scores, but those data can be easily summarized with a **frequency distribution.** The most basic frequency distribution is simply a list of each score or category and its **frequency**.

Eye Color	Frequency
Blue	10
Brown	33
Green	2
Hazel	11
Other	1
Total	57

Example of a Simple Frequency Distribution Table Using Nominal Data

Exam Score	Frequency
50	1
49	2
47	5
45	3
44	10
43	19
40	8

Exam Score	Frequency
38	3
37	1
26	1
Total	53

Example of a Simple Frequency Distribution Table Using Interval or Ratio Data

Problem With Simple Frequency Distributions

While these simple frequency distributions work well when there are not too many categories of scores, this quickly becomes less helpful when there are numerous categories. For example, assume that on a 50-point exam, every score from 0 to 50 was earned by at least one individual in the course. This would mean that 50 categories would be needed to present the data using a simple frequency distribution. This would not truly summarize the data since there would be 50 categories of scores and 53 individuals. Most categories would have a frequency of one, and only a few would have a frequency of two or more. The solution to this problem is to reduce the number of categories in the frequency distribution without losing too much information about the distribution of scores.

A Solution to the Problem: Guidelines for Grouping Scores Into Categories

While grouping scores into categories or a range of values is a good solution to increase the degree of summarization in your data, it is also important to avoid oversummarizing to the point that it obscures the pattern in the data. Fortunately, there are some common guidelines for how to create categories.

General Guidelines to Creating Categories

a. Choose between 10 and 20 categories (or intervals).

b. It is best to have a minimum of five raw scores in each category (or interval) rather than a large number of categories (or intervals) with a low frequency of scores, so use this as a guideline for the absolute number of categories or intervals to use for your data.

c. The highest score should be at the top and the lowest score at the bottom of the table.

Specific Instructions for Creating a Frequency Distribution

★ **Step 1:** Calculate the range of scores.

Range = High score – low score

★ **Step 2:** Calculate the size of your interval (or category).

Width of the interval (i) = range/number of desired intervals

Be sure to round the answer you get for i.

★ **Step 3:** Make the lower limit of the lowest interval divisible by i.

★ **Step 4:** List the intervals with the lowest scores at the bottom.

Table 3.1

17	44	43	8	17
16	13	16	36	23
30	7	37	24	28
19	22	42	39	43
33	41	38	12	34
18	14	48	50	38
25	47	35	11	9
27	15	21	49	10
32	40	26	46	29
31	5	44	32	9
45	20	50		

Raw Scores for Example Problem

Example Using Raw Scores From Table 3.1

★ **Step 1:** Calculate the range of scores.

Range = High score – low score

Range = 50 – 5 = 45

★ **Step 2:** Calculate the size of your interval (or category).

Width of the interval (i) = range/number of desired intervals

Width of the interval (i) $= \dfrac{45}{10} = 4.5$

Note: Round the solution to a whole number, so 4.5 becomes 5.

★ **Step 3:** Make the lower limit of the lowest interval divisible by i.

The lowest score in the raw data is 5, which is divisible by i (which equals 5 when we round up in this example), so the lower limit of our lowest interval will be 5.

★ **Step 4:** List the class intervals with the lowest scores at the bottom.

> 50–54
> 45–49
> 40–44
> 35–39
> 30–34
> 25–29
> 20–24
> 15–19
> 10–14
> 5–9

By adding in the frequency of the scores that fall into each interval, you get the following frequency distribution from the raw data shown in Table 3.1.

Class Intervals	Frequency
50–54	2
45–49	6
40–44	6
35–39	6
30–34	6
25–29	5
20–24	5
15–19	7
10–14	5
5–9	5
Total scores =	53

Cumulative percentage: The cumulative frequency divided by N and then the product multiplied by 100. The cumulative percentage for the highest class interval should equal 100.

Cumulative frequency: The sum of the frequency of scores starting from the lowest class interval and working up to the highest class interval. The cumulative frequency (or sum of the frequencies for each class interval) for the highest class interval should be equal to the sample size (N).

Relative frequency: The frequency of scores in that class interval divided by the total number of scores in the distribution.

More Complete Frequency Distribution Tables

Frequency distribution tables also typically include a number of other summary statistics to indicate the distribution of raw scores. Each of these statistics provides different types of information regarding the nature of the raw scores. For example, **cumulative percentage** can indicate what percentage of scores fall below a particular class interval of interest or can indicate which interval is associated with 50% of the scores above and below that point, and **cumulative frequency** is the sum of the frequency of scores starting from the lowest class interval and working up to the highest class interval. **Relative frequency** reflects how common scores are in each class interval relative to the other intervals.

Class Intervals	Frequency	Relative Frequency	Cumulative Frequency	Cumulative %
50–54	2	.0377	53	100.00
45–49	6	.1132	51	96.23
40–44	6	.1132	45	84.91
35–39	6	.1132	39	73.58
30–34	6	.1132	33	62.26
25–29	5	.0943	27	50.94
20–24	5	.0943	22	41.51
15–19	7	.1321	17	32.07
10–14	5	.0943	10	18.87
5–9	5	.0943	5	09.43
Total scores	53			

GRAPHING

Many statisticians prefer to get a first feel for their data by looking at them visually. A frequency distribution graph can quickly demonstrate the range of the scores as well as the frequency of occurrence of each score in the sample or population and can even provide a preliminary indication of the central score. Before we go over sample graphs used in statistical analyses, we'll review some basic concepts in graphing data.

Graphing Basics

Two-dimensional graphs have an X-axis (abscissa) and a Y-axis (ordinate). The X-axis is in the horizontal position, while the Y-axis is in the vertical position. It is traditional to show the score or dependent variable on the X-axis, while some characteristic of the score is shown on the Y-axis (see Figure 3.1). When the graph is a frequency distribution, the characteristic of the score shown on the Y-axis will be the frequency of that score in the distribution. In later chapters, you will often see the average score shown on the Y-axis.

General rules of thumb include (1) starting both axes at 0, (2) filling the area of the graph so that it is approximately ¾ filled, and (3) having the height at approximately ¾ of the width of the graph. If it is not feasible to have both axes begin at 0, then it is traditional to show a break on the appropriate axis with "//" to indicate that part of the scale was

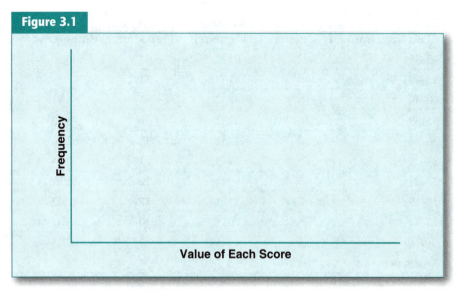

Figure 3.1

Generic Example of a Two-Dimensional Graph

dropped to avoid misinterpretation of the scale. Finally, it is important to label the *X*-axis and the *Y*-axis for the graph so that it is clear what is being represented.

Types of Frequency Graphs

Two major categories of graphs are based on whether the data on the *X*-axis are plotted as a discrete or continuous variable. For example, eye color can be categorized as a discrete variable and be plotted on the *X*-axis with the frequency of each eye color plotted on the *Y*-axis. Alternatively, scores on an exam could be considered a continuous variable (0–100) and be plotted on the *X*-axis with frequency on the *Y*-axis. We'll address how to plot discrete variables on the *X*-axis first (bar graphs) and then how to plot continuous variables on the *X*-axis (histograms, frequency curves, scatter plots).

Bar graph: Typically a two-dimensional graph wherein one dimension represents the independent variable and the other represents the dependent variable (frequency). The magnitude of the dependent variable is represented by the height of a rectangle with a uniform width (a bar).

Graphing a Discrete Variable on the X-Axis

It is common to have discrete categories of data that you plot on the *X*-axis (e.g., blue, brown, green, hazel, other) and then plot the frequency on the *Y*-axis (see Figure 3.2). This is known as a **bar graph,** and you'll see that the data on the *X*-axis are grouped in a way that is discrete rather than continuous. It can also occur with fewer discrete categories (e.g., Week 1, Week 3, and Week 5 or smoking, nonsmoking, and unknown smoking status).

Figure 3.2

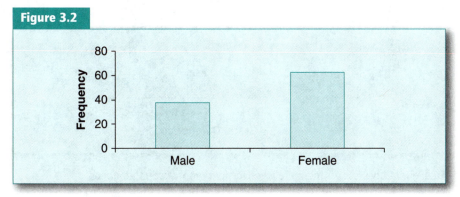

Sex Distribution in Statistics Class

Graphing a Continuous Variable on the X-Axis

There are more graphing options for a continuous variable on the *X*-axis. These include histograms (Figure 3.3), frequency curves (Figures 3.4 and 3.5), and scatter plots. **Histograms** are used to plot frequencies for continuous data and use a separate bar for each score or interval of scores.

> Histogram: The most common form is plotted by splitting the data into equal-sized intervals (*X*-axis) and then indicating the frequency of scores in each interval (*Y*-axis) with a vertical bar.

Figure 3.3

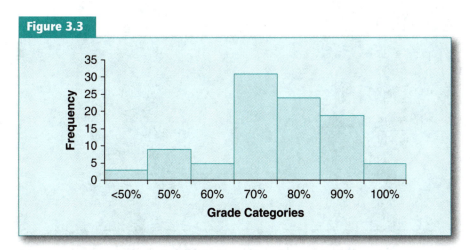

Histogram of the Distribution of Grade Categories

Figure 3.4

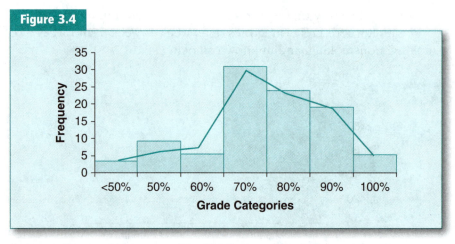

Frequency Curve and Histogram of the Distribution of Grade Categories

Frequency curve: A two-dimensional graph plotting a continuous variable on the *X*-axis and frequency on the *Y*-axis with a smoothed line. Technically, it is a smoothed version of a histogram or graphical representation of a frequency distribution.

Alternatively, you could plot the same data using a **frequency curve,** where a smoothed line is plotted to represent the data shown in the histogram (Figure 3.5). We'll discuss the properties of the frequency curve and its usefulness for inferential statistics in later chapters.

Note that each point represents one city, and it is plotted as a function of both the population size

Figure 3.5

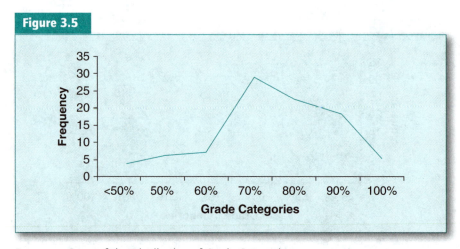

Frequency Curve of the Distribution of Grade Categories

of that city as well as the number of robberies reported in that particular period. **Scatter plots** are a useful way to compare the relationship between two variables. In our example, you can evaluate whether there appears to be a relationship between the size of the population and the frequency of robberies. Of course, you cannot necessarily determine that the relationship (if it exists) is causal (that larger population size "causes" more robberies or

Scatter plots: A graphical display of pairs of continuous data points (*X* and *Y* pairs) that is intended to reveal any potential relationship between the two variables. For example, you might plot the incidence of robbery as a function of population size for 10 cities in the United States (Figure 3.6).

Figure 3.6

	Population Size in 2005 (*X*)	Number of Robberies in 2005 (*Y*)
City 1	100,333	12
City 2	399,870	30
City 3	744,111	51
City 4	80,092	4
City 5	783,566	91
City 6	1,099,025	1,200
City 7	506,852	109
City 8	22,801	1
City 9	285,977	19
City 10	981,555	368

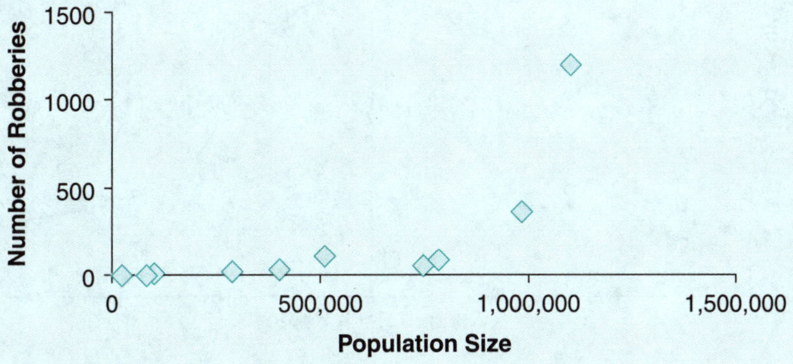

Paired Data on Population Size (*X*) and the Number of Robberies (*Y*) in 10 Cities in 2005

that more robberies "cause" larger population sizes). We will discuss this more in the chapter on correlation.

Symmetry: A distribution that is identically shaped on either side of the central point, or mirror images of one another.

Skewness: A descriptive statistical measure of whether a distribution has a higher frequency of scores on one side compared to the other side of the distribution. If there are more scores on the right-hand side of the distribution, then the distribution is negatively skewed. If there are more scores on the left-hand side of the distribution, then the distribution is positively skewed.

Shapes of Frequency Distributions

A frequency distribution is said to demonstrate **symmetry** if it is identically shaped on either side of the central score so that each side is a mirror image of the other side (see Figure 3.7). Some types of data are naturally symmetrical, while others (such as exam scores in a typical statistics class) are usually asymmetrical.

When the distribution is asymmetrical, it means that the distribution has a *negative skew* or a *positive skew,* so **skewness** is one measure of the shape of a distribution of scores (see Figure 3.8).

Frequency distributions can also differ in their "peakedness." Scores may be all clustered very close to the center of the distribution or may be very

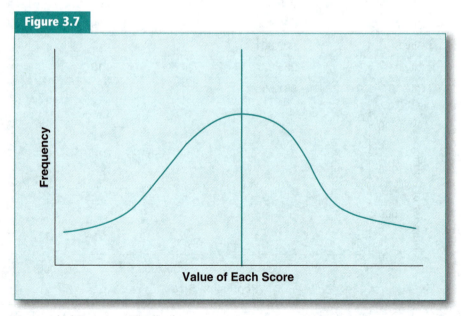

Figure 3.7

Symmetrical Frequency Distribution

Skewed Frequency Distributions

uniformly distributed about the mean. **Kurtosis** is a measure of whether the distribution is flat and spread out or peaked. A distribution can be symmetrical and flat as well as symmetrical and peaked. It is also possible for a frequency distribution to exhibit skewness as well as kurtosis (flat or peaked).

> Kurtosis: A descriptive statistical measure of whether the frequency of scores in a distribution are more or less heavily clustered around the center (peaked) as opposed to evenly distributed across the range of scores.

SUMMARY

In this chapter, you've been introduced to frequency distributions that are presented in table and graphic displays, as well as a number of other graphs used to display continuous and discrete variables. In this chapter, we've used the terms *average* and *center* to refer to the middle score. In the next chapter, we'll address specific statistical terms and techniques for representing measures of central tendency as well as those for representing variability, or the distribution of scores. Both the graphs in this chapter and the descriptive statistics we'll address in the next chapter are used to provide a summary of data, and many of them are incorporated into determining whether certain assumptions of inferential tests have been met. This material is laying the groundwork for the more sophisticated statistical analyses that you'll be doing later in the course.

CHAPTER 3 HOMEWORK

Provide a definition for the following terms.

1. What is a frequency distribution?

2. What is relative frequency?

3. What is a histogram?

4. What is a bar graph?

5. What is the ordinate?

6. What is a scatter plot?

7. What is symmetry?

What is the difference between these terms?

8. What is the difference between relative frequency and cumulative frequency?

9. What is the difference between a histogram and a bar graph?

Explain the following concepts in a short-answer format.

10. List the basic rules for creating a graph.

11. When is it best to use a scatter plot rather than a histogram?

12. If you know that a distribution is asymmetrical, then what must be true of its skewness?

13. The distribution of scores of college seniors on a third-grade spelling test would probably be _____.

    A. positively skewed

    B. negatively skewed

    C. kurtotic

    D. symmetrical

14. If biology majors took a test on theoretical concepts in social work, the distribution of scores would probably be _____.

    A. positively skewed

    B. negatively skewed

    C. kurtotic

    D. symmetrical

15. The following data, if plotted as a frequency distribution, would be
    _____.

    1 3 5 7 9 9 11 13 15 17

    A. positively skewed
    B. negatively skewed
    C. kurtotic
    D. symmetrical

16. Make a complete frequency distribution table (all columns) from the following
    raw data using $i = 10$:

66	80	89	71	80	88	82	98	83	100	72
70	64	75	79	82	88	71	85	94	93	80
77	83									

17. Make a frequency distribution graph from the data provided in Question 14.

18. Describe the shape of the distribution drawn in Question 15 using terminology
    from this chapter.

19. Rates of heart attack for urban hospitals on the East Coast are listed below.
    Generate a complete frequency distribution table, as described in this chapter.
    Use $i = 50$.

Mortality Figures for Heart Attack									
640	477	565	500	390	460	482	475	549	518
445	453	410	495	349	430	598	428	460	506
522	439	368	438	365	405	370	550	358	530
515	375	420	468	395	400	472	322	345	290
385	559	281	506	310	458	406	479	481	250

20. If English majors took a test on nursing, the distribution of scores would
    probably be _____.

    A. positively skewed
    B. negatively skewed
    C. kurtotic
    D. symmetrical

21. Create a frequency distribution from the following data. Use approximately 10 intervals.

10	14	21	35	13	18	9	13	20	15
19	21	25	30	12	16	23	30	32	17
37	8	11	31	22	33	13	6	17	25

22. The following are the test scores obtained from sociology students. With these data, create a complete frequency distribution. Use an interval width of 5.

85	98	46	72	55	92	64	78	51	82
41	73	84	67	91	55	82	56	70	65
72	56	40	89	50	63	85	90	61	46

CHAPTER 4

The Mean and Standard Deviation

I n this chapter, you will learn about some basic descriptive statistics, including the mean and standard deviation. **Descriptive statistics** are numbers that summarize your data; in other words, they are numbers that help describe your data set without giving every last raw score. Descriptive statistics can describe a large group of scores (your data) with a single number, such as a mean. We will begin the chapter with some descriptive statistics that indicate the central tendency of your sample, such as the arithmetic mean, the median, and the mode. We will then turn to measures of variability.

> Descriptive statistics: Numbers that summarize a set of data in one of four ways, including a measure of central tendency, a measure of the variability of the scores, a measure of the shape of the distribution of scores, or the size of a sample.

These concepts are important because they are used to assess whether inferential statistics can be conducted and are often included in the formulas to calculate inferential tests. Most students grasp the concepts of central tendency fairly well but struggle later in the course in understanding the differences between standard deviation and variance. Another concept that students struggle to apply later in the course is the concept of skewness and kurtosis. If you spend a little extra time on these concepts now, you'll have a much easier time later in the course. Keep in mind that we are laying the groundwork for more sophisticated statistics, so it's worth your time to learn these well so you can "ace" tests now and later. In fact, it may be beneficial to review this chapter again when you learn *t* tests later in the course.

In addition to demonstrating how to calculate these numbers manually with your calculators, we also show the output from Microsoft Excel as well as a common statistical package for the social sciences (Statistical Package for Social Sciences, SPSS). SPSS was chosen as a representative software package for statistics because it is widely used, it has a student version, and there are a large number of user-friendly manuals on how to use it.

MEASURES OF CENTRAL TENDENCY

Central tendency is the property of a distribution of scores that describes the center of the data. We often want to describe our data with a single number, and the usual choice is to describe the location of the center of the scores. This makes sense, but it turns out to be more difficult than it might otherwise initially appear. The best number to describe the "center" of the data may depend on the measurement scale of the data as well as the shape of the data. Data that are very "pretty"—that is, symmetrical—have a center that is relatively easy to describe. Data with a dramatic lack of symmetry (a distinct skew or shift to the left or right) or that have multiple peaks are not so easy to summarize in a single measure of central tendency.

The Mean

The most common measure of central tendency, or the average, is the arithmetic mean. The **mean** is equal to the sum of the scores divided by the number of scores, and it is calculated in that way whether you are calculating the mean of a sample or the mean of a population of scores.

> **Mean:** A statistic that measures the middle, or central tendency, of a set of scores and is calculated by taking the sum of the scores and dividing by the number of scores.

Formula to Calculate the Mean of a Population of Scores

$$\mu = \frac{\sum X_i}{N}.$$

Formula to Calculate the Mean of a Sample

$$\bar{X} = \frac{\sum X_i}{n}.$$

Example

A psychologist studying dyslexia recorded the spelling tests of five children who were thought to be dyslexic based on preliminary testing. The five children had the following scores on their spelling tests, which were worth 12 points each.

Calculate the mean of this sample.

Child	Score
1	6
2	8
3	10
4	4
5	11

$$\bar{X} = \frac{\sum(6 + 8 + 10 + 4 + 11)}{5} = \frac{39}{5} = 7.8$$

$$\bar{X} = 7.8$$

Properties of the Mean

The mean has certain properties or characteristics that make it an appropriate statistic to use in summarizing a data set. These include the following:

1. The mean is sensitive to extreme values.

   a. Very large or very small scores (relative to the other scores) will shift the mean so that it is less representative of the other scores in the sample or population. These extreme scores are referred to as outliers.

   b. For example, summarizing the average salary in the Seattle Metro area by using the mean would be less representative if you included the multimillion dollar salary of Bill Gates. His extremely high salary would "pull" the mean significantly upwards, and thus it would less accurately reflect the salary of most residents of the Seattle Metro area.

2. The mean is sensitive to exact values.

    c. Since the mean is calculated by summing all of the scores and dividing by the number of scores, all values equally influence the calculation of the mean. Changing any single value (data point) by any amount will change the value of the mean.

3. The sum of the deviations around the mean is equal to zero.

    d. $\sum(X_i - \bar{X}) = 0$.

    e. Thus, if you subtract the mean from every one of the scores and then sum those "deviations from the mean," the sum will equal zero. This is logical since the mean should arithmetically be the middle score.

4. The sum of the *squared* deviations is a minimum.

    f. $\sum(X_i - \bar{X})^2$.

    g. Minimum means that the result is the smallest number that you could get. If any number other than the mean was subtracted from the scores to form the squared deviations, the resulting sum of the squared deviations would be larger than if the mean were used. Thus, one definition of the mean is that number that produces a "minimum" sum of the squared deviations for any given data set. Take this characteristic at face value for now . . . we will return to it later.

5. The mean is the least subject to sampling variation.

    h. Thus, repeated samples result in relatively little variation in the value of the mean. This stability is generally a very desirable characteristic.

The Median

The **median** is the value in the middle of your scores when they are rank ordered; literally, it is the value where 50% of the scores fall below it. The term *rank order* simply means placing the scores in order from least to greatest magnitude, and the term appears later in our textbook when we

> Median: The value at which 50% of the scores fall below it.

use similar ranking to test for overlap in samples of data. The median is important because it is not sensitive to extreme scores but only to the number of scores above and below it. We can calculate the median of the spelling scores from above.

Example

★ **Step 1:** Rank the scores from smallest to largest.

$$4 \quad 6 \quad 8 \quad 10 \quad 11$$

★ **Step 2:** Find the center score.

If you have an odd number of scores, then your middle score is the median. In the above case, it is 8. If you have an even number of scores, then you must take the average of the two center scores. For example:

$$2 \quad 4 \quad 6 \quad 8 \quad 10 \quad 12$$

In this case, both 6 and 8 are the center scores. If you take the average of 6 and 8, you get 7: Median $= \dfrac{6+8}{2} = 7$.

Note that if we change the "2," say, to a "1," it would have no effect on the median: The median would still be 7. Likewise, we could change the "12" to "35" and the median would remain the same. Thus, the median is said to be relatively insensitive to exact scores.

Properties of the Median

Like the mean, the median has particular characteristics that make it an appropriate summary of central tendency in some cases. For instance:

1. The median is less sensitive to extreme scores than the mean.

   a. Because it is the center score, it will be less affected by outliers (like Bill Gates's annual income). Thus, you should always think about what the author is trying to summarize when the mean or the median is used. Someone trying to suggest that incomes are high in Seattle might summarize incomes using the mean, where a few very high incomes would highly influence the mean. The median might be a more representative measure of central tendency in this case because Bill Gates's income is just one more value on the high side of the central tendency of the data set. The median is not affected by a score that's a little bit higher than the center or a lot higher. The median is only influenced by the number of scores on either side of the center, not their actual values.

2. The median is more sensitive to sampling variation than the mean.

   b. The median will vary from sample to sample more than the mean because the middle score will only represent one score rather than an average of all scores of that particular sample.

The Mode

The **mode** is the most frequent score in a sample or population, and thus it is always a real score.

> **Mode:** The most frequent score in a set of scores.

Example

Assume that these scores are exam scores, and the total possible points are 100. Seven students took the exam.

★ **Step 1:** Rank order the scores so that duplicates are easy to identify.

50   72   72   88   89   93   100

★ **Step 2:** Determine any replicates in the sample data.

Two students earned a 72, while all other scores are unique. Thus, the mode is 72 for this sample. Note that the mode reflects a score that a student actually earned, whereas the median and mean may not. Since there are an odd number of scores, the median does reflect a true score in this case (88). However, the mean for this example is 80.57, which is not a true score.

The term *mode* most frequently pops up in describing the shape of the distribution of scores: The term *unimodal* refers to a distribution that has a single mode, a single peak, while a *bimodal* distribution has two peaks or modes, each of at least almost equal height or frequency. A *multimodal* distribution has several modes or peaks. We will find in the next chapter that unimodal distributions with certain other characteristics are particularly useful in describing our data and testing hypotheses.

Choosing the Best Measure of Central Tendency

The mean is generally considered the most appropriate measure of central tendency when the data are recorded on an interval or ratio scale and are at least approximately symmetrical. The stability of the mean in the face of sampling variation makes it highly desirable if the conditions of scale and symmetry are met. The median is more appropriate for ordinal data and for data that are not symmetrical. The mode is useful only when it is important that the measure of central tendency is a real score from the data set but is otherwise not commonly reported.

MEASURES OF VARIABILITY

Measures of variability are those that tell us how spread out the scores in a sample or population are from one another. This provides us with information that the mean does not reveal. It is possible for two samples to have the same mean but differ in their distribution of scores (their variability; see Figure 4.1). It is useful to have a measure of variability so that we can determine how far a particular score is from its mean. Students often use this as a measure of how well (or how poorly) they did on an exam. Statisticians use measures of sample variation to *estimate* the variation of a population of scores that they do not have available to them. We will describe three such measures of variability: the range, the variance, and standard deviation.

Notice that both of these distributions have a mean of 75, but the variability around the mean is not the same. In the upper distribution, the majority of the scores are between 60 and 90, but the lower distribution contains the majority of scores between 65 and 85. The bottom distribution of scores is much more closely arranged near the mean.

Range

Range: A statistic that measures the variability of a set of scores and is calculated by taking the largest score and subtracting the smallest score.

One of the easiest measures of variability to calculate is the range. The **range** is determined by subtracting the smallest score from the largest score and gives one a measure of how far apart the scores of the sample or population are distributed.

Figure 4.1

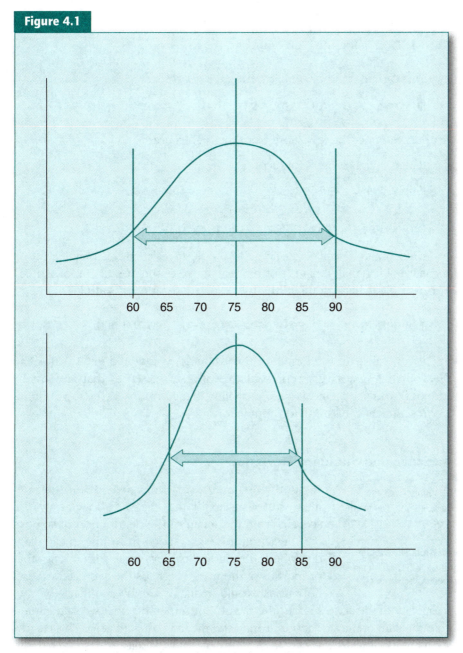

Distributions With Identical Means but Different Variabilities

Example

⭐ **Step 1:** Determine the high score and the low score.

This can be done by rank ordering the scores.

50   72   72   88   89   93   100

⭐ **Step 2:** Subtract the low score from the high score.

Range = high score − low score

Range = 100 − 50 = 50

Problems With the Range as a Measure of Variability

Unfortunately, while the range is simple to calculate, it may lead to a measure of variability that is not representative of the population of scores. For example, what if you had the following scores in your sample?

1   50   50   50   50   50   50   50   50   1,000

If you calculate the range on this example, you would get 1,000 − 1 = 999, but this is not very representative of a set of scores that are largely identical. You need a measure of variability that accounts for the distribution of *all* scores in the sample or population.

Variance and Standard Deviation

> **Variance:** The squared average deviation of a score from the mean.

> **Standard deviation:** The average deviation of a score from the mean.

Both **variance** and **standard deviation** are measures of variability that measure how far away a score is from the mean of the distribution of scores: the deviation of each score from the mean. We described deviations earlier when discussing the properties of the mean. Perhaps the average of these deviations would make a useful summary of the variability of the scores about the mean. But remember that we said earlier that the sum of the deviations around the mean is equal to zero. This clearly won't work: It will always result in a measure of variability equal to zero. But we also said that the sum of the *squared* deviations around the mean is a *minimum,* a number characteristic of that data set. So for this reason, squaring the deviations before summing them provides a better measure of variability, and generally, we calculate the *average* squared deviation by dividing the sum of the squared deviations by the number of scores (N). We might be done now, using the average squared deviations as a measure of variability (this is called the variance), but a new problem arises: The resulting squared units make the measure difficult to interpret (what are "2.34 squared seconds" as a measure of the degree of variability of some elapsed time data?). The solution to *that* problem is to take the square root of the solution after squaring the sum of the deviations. So, the standard deviation is the (deep breath!) square root of the sum of the squared deviations around the mean divided by the number of scores, or the square root of the average squared deviations. Look below to see how this idea is developed mathematically.

Remember that if you do *not* take the square root of the resulting solution and leave the answer in squared units, you have calculated the variance. That's right, the variance is left in the squared units that are more difficult to interpret descriptively, but variance turns out to be mathematically more useful than standard deviation for inferential statistics.

Sum of the Deviations Around the Mean

$$\sum(X_i - \bar{X}) = 0.$$

Sum of the Squared Deviations Around the Mean

$$\sum(X_i - \bar{X})^2$$

The sum of the squared deviations around the mean is also called the *sum of squares*. You will calculate this number often in this course as a preliminary step to calculating the standard deviation. It can be calculated the long way ($\sum(X_i - \bar{X})^2$), or it can be calculated more quickly using the following formula: $\sum x^2 - \frac{(\sum x)^2}{N}$. We prefer this second formula since the long way is more prone to errors, but we'll show you an example worked out in both ways later on. This illustrates the difference between a conceptual formula, presented for the purpose of teaching a concept, and a computational formula, much easier and more accurate for use with a calculator.

It is important to distinguish between the sum of the squared Xs and the sum of the Xs squared in using the computational method. However, if you ever get a negative sum-of-squares (SS) result, then you know that you have miscalculated. Why? Because the sum of squares represents the sum of the squared deviations: If the deviations have been squared, then they must all be positive values, and the sum of a set of positive values must be positive. The most common error is to mix up the sum of the squared Xs and the sum of the Xs squared, which will result in a negative number.

$$\text{Sum of squares (SS)} = \sum(X - \bar{X})^2 = \sum X^2 - \frac{\left(\sum X\right)^2}{N}.$$

FYI: HOW TO USE YOUR CALCULATOR

It is difficult to give you instructions on how to use your calculator since they vary so much, but there are excellent resources on the Internet to help you with this task. It is very important to take the time to learn how to enter data and to use the memory storage function on your calculator. Look up the instruction manual for your calculator online if you don't have it handy.

In general, these are the steps to calculate statistics such as sums of squares and standard deviation:

1. Enter in the raw data or values by punching in the value and some memory key for your calculator.
2. Read out the results by pressing some combination of keys to get to the statistical symbol of interest on your calculator.

Be aware that some keys may contain multiple available statistical functions, and you may have to do the equivalent of a shift key press to access the desired symbol.

Learning how to do this will save you significant time and errors on your homework and exams.

Formula for the Variance of a Population of Scores

The variance is the sum of the squared deviations around the mean divided by the number of scores. Mathematically, that is represented with the following notation:

$$\sigma^2 = \frac{\sum(X_i - \mu)^2}{N} \text{ or, more easily,}$$

$$\sigma^2 = \frac{\sum X^2 - \frac{\left(\sum X\right)^2}{N}}{N}.$$

The population variance is represented by sigma (σ^2). Note also that we have used the mean of the population (μ) in this formula since we are calculating the population variance.

If you substitute the sum-of-squares symbol for the sum of the squared deviations, you can rewrite the formula:

$$\sigma^2 = \frac{SS}{N}.$$

Formula for the Variance of a Sample of Scores

$$s^2 = \frac{\sum(X_i - \bar{X})^2}{n - 1} \text{ or again, more easily,}$$

$$s^2 = \frac{\sum X^2 - \frac{(\sum x)^2}{n}}{n - 1}.$$

If you substitute the sum-of-squares symbol for the sum of the squared deviations, you can rewrite the formula:

$$s^2 = \frac{SS}{n - 1}.$$

Formula for the Standard Deviation of a Population of Scores

The standard deviation is the square root of the sum of the squared deviations around the mean divided by the number of scores. Mathematically, that is represented with the following notation:

$$\sigma = \sqrt{\frac{\sum(X - \mu)^2}{N}} \text{ or the computational version } \sigma = \sqrt{\frac{\sum X^2 - \frac{(\sum x)^2}{N}}{N}}.$$

This formula is similar to the formula for the population variance except that the symbol for standard deviation demonstrates that the units are no longer squared (σ) because we have taken the square root of the solution.

If you substitute the sum-of-squares symbol for the sum of the squared deviations, you can rewrite the formula:

$$\sigma = \sqrt{\frac{SS}{N}}.$$

Taking the square root of the sum of the squared deviations around the mean allows us to come up with a measure of variability that is in the same units as our scores. If we left them in squared units, they would look artificially large, but if we did not square the deviations, we would always get zero, which is unlikely to represent the true variability of the scores. Thus, mathematically, we have solved our dilemma of finding a measure to accurately represent all scores and provide a minimum measure of variability. The measure is divided by the size of the population to adjust the total deviation by the number of scores in the population. This is what makes our measure an *average* deviation, which is more commonly called the standard deviation.

Formula for the Standard Deviation of a Sample of Scores

$$s = \sqrt{\frac{\sum(X_i - \bar{X})^2}{n - 1}} \text{ or the computational version } \sqrt{\frac{\sum X^2 - \frac{(\sum X)^2}{n}}{n - 1}}.$$

This formula is similar to the formula for the sample variance except that we have taken the square root of the solution.

If you substitute the sum-of-squares symbol for the sum of the squared deviations, you can rewrite the formula:

$$s = \sqrt{\frac{SS}{n - 1}}.$$

The standard deviation formula is altered slightly to apply it to samples rather than populations. The standard deviation of a sample is represented with the symbol s, and we use the sample mean notation ($\bar{X}$) as well. The more significant change is that we are now dividing by $n - 1$. We use the small n to indicate it is the size of a sample (rather than a population), and we subtract 1 from n.

Why do we use $n - 1$ for a sample but not a population?

Mathematicians have found that when you estimate the population standard deviation from a sample standard deviation, it tends to underestimate the true variability of the population by a little. This is important because one of our primary goals in statistics is to make inferences from samples to the underlying populations that the samples represent. It turns out that if you subtract 1 from your sample size, it provides an adjustment that ensures that your estimate of the population standard deviation is not too small. Thus, the sample standard deviation is considered a *biased* estimate of the population standard deviation, but the sample mean is an *unbiased* estimate of the population mean and requires no correction.

The term $n - 1$ is referred to as the "degrees of freedom." We will revisit the concept of degrees of freedom and explain the term in a more formal way in a later chapter.

Example (Calculated the Long Way [Conceptual] and the "Fast Way" [Computational])

Determine the sample standard deviation and variance using the conceptual formula and then the faster calculation technique.

★ **Step 1 (conceptual formula):** Determine the mean of this population of scores.

$$\bar{X} = \frac{\sum(3 + 4 + 4 + 5 + 5 + 5 + 5 + 6 + 6 + 7)}{10} = \frac{50}{10} = 5$$

★ **Step 2 (conceptual formula):** Subtract each score from the mean and then square the difference (deviation).

Score	Sample Mean	Difference (Deviation)	Squared Difference (Squared Deviation)
3	5	−2	4
4	5	−1	1
4	5	−1	1
5	5	0	0
5	5	0	0
5	5	0	0
5	5	0	0
6	5	1	1
6	5	1	1
7	5	2	4
Sum		**0**	**12**

★ **Step 3 (conceptual formula):** Solve for the variance using the formula.

$$s^2 = \frac{\sum(X_i - \bar{X})^2}{n - 1}.$$

$$s^2 = \frac{12}{9} = 1.333333.$$

★ **Step 4 (conceptual formula):** Solve for the standard deviation using the formula.

$$s = \sqrt{\frac{\sum(X_i - \bar{X})^2}{n - 1}}.$$

$$s = \sqrt{\frac{12}{9}} = \sqrt{1.333333} = 1.154701.$$

★ **Step 1 (computational formula):** Calculate $\sum X^2$ and $\sum X$.

Punch your data (10 scores) into your calculator using the statistical mode and retrieve $\sum X^2$ and $\sum X$ to insert into the formula or square each score individually (shown in the table below).

$$SS = \sum X^2 - \frac{(\sum X)^2}{n}.$$

$$SS = 262 - \frac{(50)^2}{10} = 262 - \frac{2500}{10} = 262 - 250 = 12.$$

Notice how we get the same value for SS (12) using this formula as we did for the conceptual formula above.

★ **Step 2 (computational formula):** Solve for the variance using the formula.

$$s^2 = \frac{SS}{n - 1}.$$

$$s^2 = \frac{12}{9} = 1.333333.$$

★ **Step 3 (computational formula):** Solve for the standard deviation using the formula.

$$s = \sqrt{\frac{SS}{n-1}}.$$

$$s = \sqrt{\frac{12}{9}} = = \sqrt{1.333333} = 1.154701 .$$

Score	Squared Score
3	9
4	16
4	16
5	25
5	25
5	25
5	25
6	36
6	36
7	49
Sum of X = 50	**Sum of X^2 = 262**

Score	
Mean	**5**
Standard Error	0.365148
Median	5
Mode	5
Standard Deviation	**1.154701**
Sample Variance	**1.333333**
Kurtosis	0.080357
Skewness	0
Range	4
Minimum	3
Maximum	7
Sum	50
Count	**10**

Results if You Use Excel to Calculate Sample Descriptive Statistics

The output on page 57 shows what you get from Excel when you type in the data for this example. Note that Excel provides all of the descriptive statistics that we have discussed so far as well as some we discuss at the end of this chapter (skewness and kurtosis). We have bolded the numbers that are comparable to the numbers that we just manually calculated.

Descriptive Statistics

	N	Range	Minimum	Maximum	Sum	Mean		Std.	Variance	Skewness		Kurtosis	
	Statistic	Statistic	Statistic	Statistic	Statistic	Statistic	Std. Error	Statistic	Statistic	Statistic	Std. Error	Statistic	Std. Error
SCORE	10	4.00	3.00	7.00	50.00	5.0000	.3651	1.1547	1.333	.000	.687	.080	1.334
Valid N (listwise)	10												

Results if You Use SPSS to Calculate the Sample Descriptive Statistics
(and Select All Options for Descriptive Statistics)

Note that the output from SPSS contains the same information that you saw with the Excel output, but you also have some differences. For example, SPSS includes the standard error of the skewness and the standard error of the kurtosis, which we did not get with Excel. The meaning of these numbers will be clearer to you after you've learned about the standard error of the mean, which both SPSS and Excel produce, but we have not yet reviewed this concept.

How would these descriptive statistics be reported in a scientific journal article?

If the journal required American Psychological Association (APA) format, these descriptive statistics might be reported in a table such as Table 4.1.

Table 4.1

Mean	Standard Deviation	Variance
5	1.1	1.3

Descriptive Statistics for "Score"

Properties of the Standard Deviation

Like the measures of central tendency, the standard deviation has certain properties or characteristics:

1. The standard deviation is a measure of variability of scores around the mean.

2. The standard deviation is sensitive to each score in the distribution. That is, since every score is included in the calculation of the standard deviation, if one score is changed at all, the value of the standard deviation is changed, unlike our first measure of variability, the range.

3. Like the mean, the standard deviation is not strongly influenced by sampling variation. If you (conceptually) take repeated samples of a population and calculate the standard deviation for each sample, the standard deviation will be relatively similar in each sample. That is, there will be very little variation in the estimates of standard deviation that is simply due to sampling choices.

4. The standard deviation—or, more specifically, variance—can be manipulated algebraically, which makes it useful for inferential statistics. Standard deviations cannot be added and averaged where variances can be. Later, when we will be working with multiple samples and therefore multiple estimates of variation, this property will become important.

SYMMETRY

Many statisticians prefer to get a first feel for their data by looking at them visually rather than with the descriptive statistics we have already discussed in this chapter. It is important to know what your data look like, to be able to later determine the appropriate ways to use statistics to test hypotheses. Central tendency is important (e.g., where is the center of your data relative to that of a control group?), as is the degree of variability. Another important consideration is **symmetry:** the degree to which the data are distributed equally above and below the center. The degree of symmetry will be important in assumptions for testing hypotheses.

> **Symmetry:** A distribution that is identically shaped on either side of the mean, or mirror images of one another.

A frequency distribution graph can quickly demonstrate the range of the scores as well as the frequency of occurrence of each score in the sample or population and can even provide a preliminary indication of the central score. If a distribution of scores is unimodal (has only one mode) and symmetrical (identical on each side of the mean), then the mean = median = mode. This is a way to tell, from the descriptive statistics, whether your data are unimodal and symmetrical, an important characteristic in a later chapter. Note that in the following Excel example, the mean, median, and mode all equal 5.

Score	
Mean	**5**
Standard Error	0.365148
Median	**5**
Mode	**5**
Standard Deviation	1.154701
Sample Variance	1.333333
Kurtosis	0.080357
Skewness	0
Range	4
Minimum	3
Maximum	7
Sum	50
Count	10

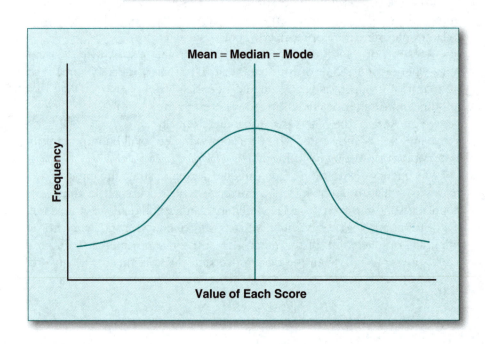

SKEWNESS AND KURTOSIS AS
QUANTIFIABLE DESCRIPTIVE STATISTICS

In a previous chapter, we reviewed skewness and kurtosis conceptually in terms of the "central value," and now we define it in terms of the measures of central tendency described in this chapter. In addition, we can now discuss skewness and kurtosis as descriptive statistics and apply numerical values to the degree of skewness and kurtosis shown in a distribution.

If the mean is less than the median, then the distribution has a *negative skew,* but if the mean is greater than the median, then the distribution has a *positive skew.* So **skewness** is one measure of the shape of a distribution of scores. There is a statistical measure of skewness, as there are central tendency and variability, which is frequently provided with statistical software output. The measure of skewness ranges from – infinity to infinity, with zero representing a symmetrical (nonskewed) distribution. A positive value of skewness means that the long tail is to the right, or toward higher values, and that the majority of scores are below the mean. A negative value of skewness means the opposite, that the tail is toward the smaller values and that the bulk of scores are above the mean.

> **Skewness:** A descriptive statistical measure of whether a distribution has a higher frequency of scores on one side of the mean compared to the other side of the mean. A distribution with a skewness = 0 is normally distributed and not skewed.

Distributions can also differ in their "peakedness." Scores may be all clustered very close to the center of the distribution (the mean) or may be very uniformly distributed about the mean. **Kurtosis** is a measure of whether the distribution is flat and spread out or peaked. This is more difficult to determine since the mean and median are identical in each case. Like skewness, there is a statistical summary statistic called kurtosis, which ranges from –infinity to infinity with an intermediate value of zero. Positive kurtosis is indicative of peakedness, and a negative value of kurtosis indicates a spread-out or flatter distribution. Note that in the following Excel example, the distribution is symmetrical (because skewness = 0 and we know that the mean, median, and mode are equal), but the distribution is *slightly* kurtotic (kurtosis = 0.080357).

> **Kurtosis:** A descriptive statistical measure of whether the frequency of scores in a distribution is more or less heavily clustered around the mean compared to a normal distribution. A distribution with a kurtosis = 0 is normally distributed and not kurtotic.

Score	
Mean	**5**
Standard Error	0.365148

(Continued)

(Continued)

Score	
Median	**5**
Mode	**5**
Standard Deviation	1.154701
Sample Variance	1.333333
Kurtosis	**0.080357**
Skewness	**0**
Range	4
Minimum	3
Maximum	7
Sum	50
Count	10

SUMMARY

You have now completed an introduction to descriptive statistics. Many of these statistics are used to provide a summary of data, and many of them also are incorporated in determining whether certain assumptions of inferential tests have been met. We have enclosed a printout from Microsoft Excel demonstrating the descriptive summary statistics included in their package. Note that the measures of central tendency and many of the measures of variability you have learned are indicated in Excel.

Excel Step-by-Step: Step-by-Step Instructions for Using Microsoft Excel 2003 or 2007 to Run Descriptive Statistics

1. Your first step will be to open Microsoft Excel and type the raw data into a spreadsheet (data listed on next page). It is helpful to type the column headers so that your output will be labeled later.

Caffeine First Times	Caffeine Second Times
30	18
32	15
30	17
33	20
32	18

2. Once your data are entered, you will need to calculate the descriptive statistics on these data.

For Excel 2003: To get all of the descriptive statistics you will need, you can go to the built-in data analysis function. Not only will you get the mean, median, mode, standard deviation, and so on, but you'll get them all in one step! You'll find the option under the "Tools" menu, and at the bottom of the list that pops up, you should see "Data Analysis."

If you do not see the "Data Analysis" option under the Tools menu, select "Add Ins" under the Tool menu. Check the box next to "Analysis TookPak" and click OK. Follow any further instructions that the computer gives you.

For Excel 2007: To get to the data analysis option, click on the "DATA" tab, and the "Data Analysis" tool will be in the "Analysis" section to the far right of the screen. Once you find the Data Analysis tool, the rest of the instructions are the same.

3. After you click on "Data Analysis," a list of possible statistical tests will pop up, and you should select "Descriptive Statistics." Once you've done this, you'll have to use your mouse to highlight the cells you want Excel to use in calculating means, variances, and so on. Be sure to click the "Summary Statistics" box.

Here is example output of these descriptive statistics from Excel:

Control Group		Alcohol Group	
Mean	2.7	Mean	4
Standard Error	0.395811403	Standard Error	0.394405319
Median	3	Median	4
Mode	3	Mode	5
Standard Deviation	1.251665557	Standard Deviation	1.247219129

(Continued)

(Continued)

Control Group		Alcohol Group	
Sample Variance	1.566666667	Sample Variance	1.555555556
Kurtosis	−0.065963914	Kurtosis	−0.91180758
Skewness	0.280477343	Skewness	−6.16791E-17
Range	4	Range	4
Minimum	1	Minimum	2
Maximum	5	Maximum	6
Sum	27	Sum	40
Count	10	Count	10

SPSS Step-by-Step: Step-by-Step Instructions for Using SPSS to Run Descriptive Statistics

1. Your first step will be to open SPSS and select the option that allows you to type in new data.

2. This will open a page called "Variable View." To confirm that, look at the tab at the bottom left of the page. There should be two tabs, and one will say "Variable View" (the one you are in now), and the other will say "Data View."

3. Now you need to establish your variables for SPSS. Make a variable for your control group by typing in the word *control* in the first box of row 1. Now name another variable for your experimental group, called "exp," and type that into box 1 in row 2. By default, SPSS will consider each of these variables to be numeric, and for these purposes, all of the default codes will work perfectly. However, keep in mind that this is where you can change some of your options to allow for alphabetical data, define coding in your variables, and so on.

4. Click on the Data View tab now. You should see that the variable names you entered in Variable View have now appeared at the top of this spreadsheet. Now you can enter your raw data:

Control	Exp
30	18
32	15
30	17
33	20
32	18

5. From the SPSS menu, you should now select "Analyze," then "Descriptive Statistics," and finally "Descriptives." This will open up a new pop-up window with your variables listed on the left-hand side. Select each variable and use the arrow in the middle of the pop-up to move each variable to the right-hand side of the pop-up box.

6. Click OK, and you should get some basic descriptive statistics, including mean, standard deviation, minimum and maximum scores, and your sample size, or n. If the descriptive statistics you want are not appearing here, keep in mind that selecting "options" on the pop-up menu will enable you to customize what descriptive statistics you would like calculated.

CHAPTER 4 HOMEWORK

Provide a definition for the following terms.

1. What is the mean?

2. What is the median?

3. What is the mode?

4. What is unimodal?

5. What is symmetrical?

What is the difference between these terms?

6. What is the difference between the standard deviation and the variance?

7. What is the difference between a positive and a negative skew?

Explain the following concepts in a short-answer format.

8. What do we mean when we say that the mean is unbiased but the standard deviation is biased?

9. Why is the sum of the deviations around the mean equal to zero?

10. Why do we divide the sum of the squared deviations by the sample size to calculate variance?

11. The following sample of data represents the amount of time spent by northwestern crows searching for food in a randomly chosen 30-minute observation period, in minutes.

    23   14   18   20   21   29   18   20

    A. Calculate the mean, sums of squares, and standard deviation of these data as if they are the only data that you care about, and you are not generalizing to another population.

    B. Estimate the population mean and standard deviation.

    C. Estimate the population variance.

12. A nursing researcher has recorded the age at which infants in an intensive care nursery achieve a milestone of reflex development. The data, in days, are below.

    28   25   26   29   26   25   26   27

    The researcher would like to draw conclusions about the performance of infants cared for in this nursery: Produce the appropriate numbers for mean and standard deviation.

13. A sociology researcher is interested in the number of vampires killed during *Buffy the Vampire Slayer* (TV show). Out of the total population of *Buffy* shows that have aired to date, the researcher randomly selects 12 episodes and records the number of vampires killed in each episode (numbers below).

| 19 | 28 | 7 | 35 | 9 | 13 |

| 88 | 17 | 2 | 21 | 19 | 11 |

Calculate the following based on the above example:

Mean =

Median =

Mode =

n =

Standard deviation =

14. The length of hospital stay (in days) was recorded for nine patients in the burn unit of a local hospital. The following nine patients are a random sample of all the patients treated at this burn unit. Assume the scores are normally distributed. Find the mean, standard deviation, and variance of these scores.

| 81 | 71 | 63 | 69 | 66 | 77 | 68 | 73 | 75 |

A. Mean =

B. Standard deviation =

C. Variance =

Use the following story problem and data set (provided below) to answer homework Questions 15 to 23.

A researcher wants to determine how long it takes on average for a voter to select a candidate at the voting booth. Out of his entire population of voters, he randomly selects 10 voters and collects the following data (in seconds):

15

13

19

7

9

11

13

8

10

12

Calculate the following:

15. Mean =

16. Median =

17. Mode =

18. $N =$

19. $\sum X =$

20. $\left(\sum X\right)^2 =$

21. $\sum X^2 =$

22. SS =

23. Estimate the population standard deviation =

The following homework questions should be answered with the online data set provided for this chapter via the textbook's website.

24. Produce a table of descriptive statistics using Microsoft Excel.

25. Interpret the descriptive statistics produced by Excel. Are your data skewed? If so, how?

26. Interpret the descriptive statistics produced by Excel. Are your data kurtotic? If so, how?

27. How are the standard deviation and the variance related to one another? Document this difference using the numbers in your Excel output.

CHAPTER 5

The Normal Distribution, Standardized Scores, and Probability

What you learn in this chapter will help explain how standard achievement tests (e.g., Scholastic Aptitude Test or SAT) are interpreted so that comparisons can be made between students. Granted, you are already in college, so SAT scores are likely not so important to you, but are you thinking about a higher degree? If so, you will encounter similar standardized scores for graduate school (e.g., Graduate Record Exam or GRE).

In this chapter, you will learn about the normal distribution, which you will use for the remainder of the course, and you will get your first introduction to calculating a statistic and then determining the probability of obtaining that value if only chance factors are at work. The math in this chapter is basically a formula that uses the numbers you learned to calculate in Chapter 4 (mean and standard deviation). So, in that sense, this chapter builds on Chapter 4.

While you have to learn the math to solve the problems presented in this chapter, the concepts in this chapter are even more critical to your ability to understand the field of statistics. Specifically, understanding that we calculate an inferential statistic and then compare the likelihood of that specific number occurring simply due to random chance is the basis of inferential statistics. The section on probability in this chapter is written to lay a foundation for you to understand this concept. Our textbook focuses more on how probability is used in statistics and less on how to calculate various probability problem sets.

NORMAL CURVE

The normal curve is a theoretical distribution with specific mathematical characteristics and properties. It is common for basic biological data, such as weight and height, to be distributed normally. In fact, most popular inference tests rely on the data that you analyze being normally distributed. Thus, we would like our data to be normally distributed or to be close enough to a normal distribution to allow us to draw inferences.

The normal curve is mathematically defined, but its formula is not necessary to your understanding of these materials. Nevertheless, it is important that you recognize its shape and properties.

Characteristics of the normal distribution (refer to Figure 5.1):

1. The distribution provides a description of the frequency of occurrence of each possible score (as does any frequency distribution).

2. This frequency distribution is symmetrical around the mean (each side is a mirror image of the other side) and unimodal.

3. The curve has inflection points (points at which the curve changes curvature direction) that are associated with the scores that are one standard deviation above the mean and one standard deviation below the mean.

4. The total area under the curve is equal to 1.00. If you think of this as a percentage, then 50% of the area lies above the mean and 50%

Figure 5.1

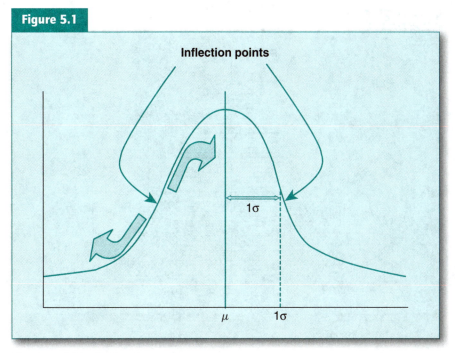

The Normal Distribution Curve

of the area lies below the mean, and the total area is equal to 100%. In fact, we know the area of the curve associated with each point on the horizontal line (*X*-axis) and can look up that area in a table (see Table A). This is displayed graphically in Figure 5.2: This figure displays the area under the curve for each standard deviation unit from the mean. Note, for instance, that if you sum all of the areas, the total is 100%. If you sum the areas in the left half, the total comes to 50%. The area under the curve between the mean and a score that is one standard deviation unit greater than the mean (whatever that score would be, given the mean and standard deviation of the data) is 34.13%.

5. Theoretically, the tails are asymptotic, meaning that both the positive and the negative tails approach the *X*-axis as the positive and negative scores get more extreme, but the tails never actually reach zero frequency (the *X*-axis).

Figure 5.2

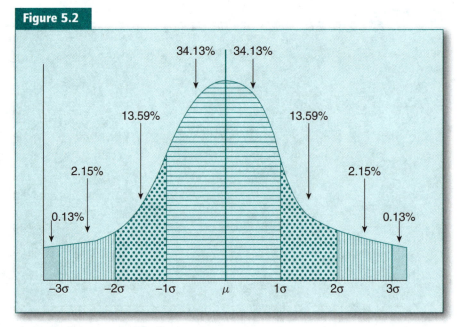

The Area of the Normal Distribution

STANDARDIZED SCORES

Standardized scores show the relative status of a score in the distribution. Knowing that someone scored a 60 on an exam means little without knowing the distribution of scores and the mean. After all, a 60 could be the low score or the high score or fall anywhere between the low and high scores. It's also important to keep in mind how far apart the scores in the distribution are from one another on average. Let's say that you scored a 60 on each of the following exams. Can you rank your performance on the three exams? It's fairly clear that Exam A was your best given the mean, but what about Exams B and C? We can place the three on the same scale and directly compare them by *standardizing* the scores. The result is called a *z* score.

Exam A: $\mu = 30$ $\sigma = 10$

Exam B: $\mu = 65$ $\sigma = 20$

Exam C: $\mu = 65$ $\sigma = 2$

Formula for a *z* Score, or Standard Score Calculation

$$z = \frac{X - \mu}{\sigma}.$$

Notice that you subtract the population mean from the score and then divide by the population standard deviation. This formula converts scores on any scale to the same scale so that they can be directly compared. It places all scores onto a scale that is relative to the position of the center of the individual distributions (the mean) and relative to the distribution of the scores (the standard deviation). So the *z* score is the original score(s) converted to a score that represents the number of standard deviation units that each raw score is from its own mean. Let's work through the three exam examples to demonstrate the *z* score principle.

 Step 1: Apply the formula to each exam score, mean, and standard deviation.

Exam A: $z = \dfrac{60 - 30}{10} = \dfrac{30}{10} = 3.00.$

Exam B: $z = \dfrac{60 - 65}{20} = \dfrac{-5}{20} = -0.25.$

Exam C: $z = \dfrac{60 - 65}{2} = \dfrac{-5}{2} = -2.50.$

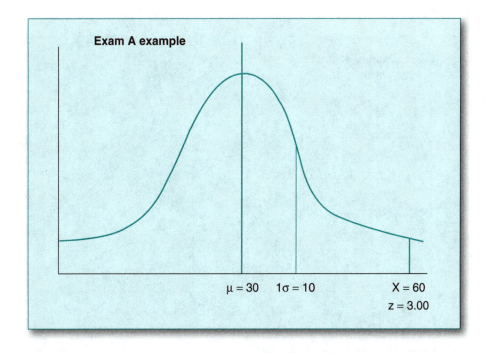

Exam A example

$\mu = 30$ $1\sigma = 10$ $X = 60$
$z = 3.00$

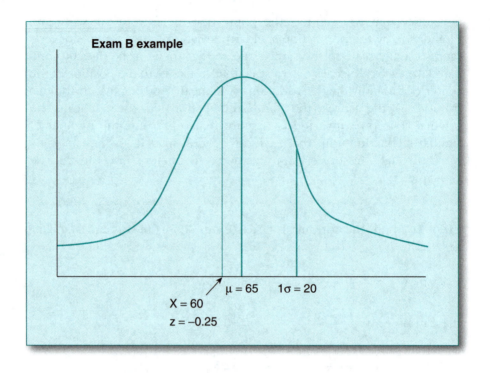

Exam B example

$\mu = 65$ $1\sigma = 20$

$X = 60$
$z = -0.25$

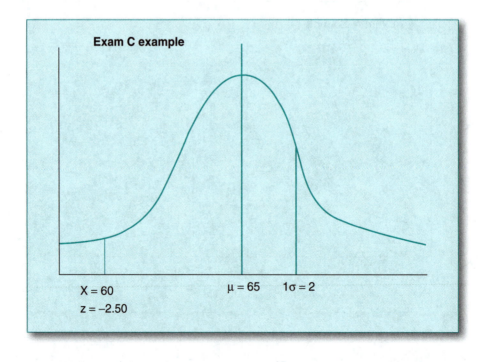

Exam C example

$X = 60$
$z = -2.50$

$\mu = 65$ $1\sigma = 2$

Step 2: Evaluate the z scores based on the sign and value of the z score.

Even without going to Table A, you could evaluate these three z scores. Because the z score for Exam A is positive, you know that it is above the mean, or better than average. In fact, it is three standard deviations above the mean, so it is far above the mean. Likewise, the z scores for Exams B and C are below the mean, and you know that is the case because they are negative z scores. Having a larger negative z score (as in Exam C) indicates a performance far below the mean, while Exam B is slightly below the mean in comparison.

 Step 3: Evaluate the z scores based on the normal distribution curve (Table A).

One of the most important characteristics of a normal curve is that we know the area under the curve for each value on the X-axis. We have converted the exam scores into standardized scores on a normal curve. Thus, we can look up the z values that we have calculated to determine the area under the curve for each of our three values. Because the normal distribution is symmetrical, we can look up the positive numbers, and the area under the curve would be identical on the other side. Column C in Table A gives you the area under the curve from your z score into the nearest tail. Column B gives you the area between that particular z score and the mean. Column A is each z score. Begin by looking in column A until you find the z score of interest. Whether you use column B or C to find the area depends a bit on the location of the z score. It is best to draw yourself a quick picture when solving these problems so that you pick the correct column and calculate the area appropriately.

Exam A: $z = 3.00$ (from Table A, column C); the area beyond z is 0.0013. Thus, if you change this to a percentage (multiply by 100), you find that only 0.13% of all scores are this far *above the mean* or further (more extreme than this score).

Col A	Col B	Col C
.	.	.
.	.	.
.	.	.
2.99	.4986	.0014
3.00	.4987	.0013
3.01	.4987	.0013
.	.	.
.	.	.
.	.	.

Table A (Reproduced)

Exam B: $z = -0.25$ (from Table A, column C); the area below z is 0.4013. Thus, if you change this to a percentage (multiply by 100), you find that 40.13% of all scores are this far *below the mean* or further below the mean (more extreme than this score).

Col A	Col B	Col C
.	.	.
.	.	.
.	.	.
0.24	.0948	.4052
0.25	.0987	.4013
0.26	.1026	.3974
.	.	.
.	.	.
.	.	.

Table A (Reproduced)

Exam C: $z = -2.50$ (from Table A, column C); the area below z is 0.0062. Thus, if you change this to a percentage (multiply by 100), you find that only 0.62% of all scores are this far *below the mean* or further below the mean (more extreme than this score).

Col A	Col B	Col C
.	.	.
.	.	.
.	.	.
2.49	.4936	.0064
2.50	.4938	.0062
2.51	.4940	.0060
.	.	.
.	.	.
.	.	.

Table A (Reproduced)

Notice that in all three examples, we used column C. This is because column C is the most intuitive to use, but we will demonstrate some problems that require the use of column B. If you refer back to Table A with a *z* score of 3.00, notice that column B indicates that the area between a *z* score of 3.00 and the mean is 0.4987 or 49.87%. If you add the area from column C to that (0.0013), you should get .5000 or 50%. This is correct because the two sides of a normal distribution are symmetrical (each contain ½ of the total area), and the total area under the curve is 1.00 or 100%. We'll go through another example to clarify this point.

Example

John scored 43 out of 60 possible points on his first introductory sociology exam. If the exam mean was 30 and the standard deviation was 8, what proportion of students scored below John and what proportion scored above John?

 Step 1: Apply the formula to John's exam score to calculate his *z* score.

$$z = \frac{X - \mu}{\sigma}.$$

$$z = \frac{43 - 30}{8} = \frac{13}{8} = 1.625.$$

 Step 2: Evaluate John's *z* score based on the normal distribution curve (Table A).

Look for *z* = 1.625 in column A. You'll find a *z* score = 1.62 and 1.63 but not 1.625. You can average the area under the curve between these two scores or just pick one to get an approximate area. If you average the area associated with a *z* = 1.62 in column C (area = 0.0526) and the area in column C associated with *z* = 1.63 (area = 0.0516), you get an area beyond your *z* score of 0.0521. Thus, 0.0521 or 5.21% of the students scored as well or better than John on the exam. How many scored below John? If you apply the fact that the total area under the curve is 1.00 or 100%, then you can just subtract the column C area from this number to get the number of students that scored below John (1.00 − 0.0521 = 0.9479 or 94.9%). You could also use column B to solve this problem. Column B indicates that the area between John's *z* score of 1.625 and the mean can be averaged as 0.4479. If you add .50 or 50% to that number, then you have the total area below John. Again, we highly recommend drawing yourself a quick graph to ensure that you do not get columns B and C confused. It is as easy as confusing right and left!

Col A	Col B	Col C
.	.	.
.	.	.
1.61	.4463	.0537
1.62	.4474	.0526
1.63	.4484	.0516
1.64	.4495	.0505
.	.	.
.	.	.
.	.	.

Table A (Reproduced)

An Additional (More Challenging) Example

The testing bureau reports that the mean for the population of GRE scores is 500 with a standard deviation of 90. The scores are normally distributed. What proportion of scores lies between the scores of 460 and 600?

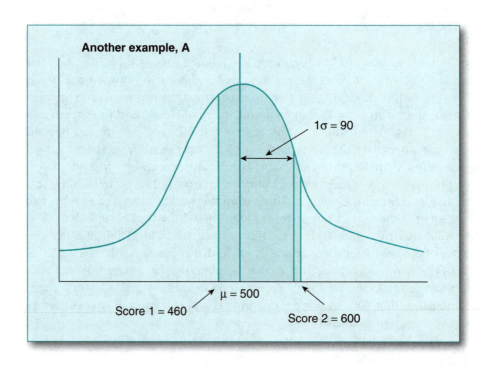

Step 1: Apply the z score formula to each exam score to calculate each z score.

$$z = \frac{X - \mu}{\sigma}.$$

$$z = \frac{460 - 500}{90} = \frac{-40}{90} = -0.4444.$$

$$z = \frac{600 - 500}{90} = \frac{100}{90} = 1.1111.$$

 Step 2: Evaluate each z score based on the normal distribution curve (Table A).

Because you want to know the area between each z score rather than beyond each z score (into the tails of the distribution), this is an ideal use of column B areas.

$z = -0.4444$ (from Table A, column B); the area between z and the mean is approximately 0.17. Thus, if you change this to a percentage (multiply by 100), you find that 17% of all scores fall between this score and the mean.

$z = 1.1111$ (from Table A, column B); the area between z and the mean is approximately 0.3665. Thus, if you change this to a percentage (multiply by 100), you find that 36.65% of all scores fall between this score and the mean.

 Step 3: Add up the areas to get the total area between the scores.

Area between $z = -0.4444$ + area between $z = 1.111$ = total area between z scores 0.1700 + .3665 = .5365 or 53.65% of all scores fall between GRE scores of 460 and 600.

What if you know the area but want to find the score?

Let's say that you know the area but want to find the score associated with that area. This score is called a percentile. You might want to know what score is associated with an area to the left (or less than the score) of 0.75 or 75% (the 75th percentile), or you might want to know what score on the GRE you need to obtain to be in the top 10% (or above 90% of the scores: the 90th percentile) of all applicants to graduate school. The z score formula can answer these questions as well; you just need to rearrange the formula to solve for X rather than for z.

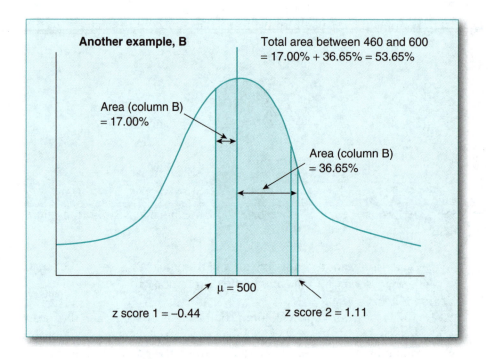

Another example, B

Total area between 460 and 600
= 17.00% + 36.65% = 53.65%

Area (column B)
= 17.00%

Area (column B)
= 36.65%

$\mu = 500$

z score 1 = −0.44

z score 2 = 1.11

The z Score Formula

$$z = \frac{X - \mu}{\sigma}.$$

Rearrange the Formula to Solve for a Score (X)

$$X = \mu + \sigma(z).$$

Example

If the mean of an exam is 60 and the standard deviation is 20, what score is associated with the 75th percentile (75% of the scores being less than this score)?

 Step 1: Apply the rearranged formula to these parameters.

$$X = \mu + \sigma(z).$$
$$X = 60 + 20(z).$$

Look in Table A, column C to find the area in column C that is associated with 0.25 (25% above or alternatively 75% of the scores below this score). You could also look for 0.25 in column B. You should find that $z = 0.67$.

$$X = 60 + 20(0.67) = 73.4.$$

Step 2: Evaluate your calculations.

A score of 73.4 is the score where 75% score below or 25% score above this value.

PROBABILITY

Probability (or its symbol, p) is a number between 0 and 1 that indicates the likelihood of an event occurring due to chance factors alone: We will define probability more formally below. We can combine this concept and that of normally distributed variables by using z scores. Instead of asking for the area under the curve or the percentage of individuals falling above or below the z score, we can ask for the probability of obtaining a specific z score. For example, what is the *probability* of obtaining a score of 98 or better on an exam where the mean is 60 and the standard deviation is 20? This should sound familiar: It's just a z score problem with the word *probability* substituted for *area* or *proportion.*

 Step 1: Insert numbers into the z score formula and calculate z obtained.

$$z = \frac{98 - 60}{20} = 1.9.$$

 Step 2: Evaluate the statistic.

From Table A, column C, you find that the area under the curve or the likelihood of obtaining a z score of 1.9 or higher is 0.0287 or 2.87%. Using our old terminology, we would say that 2.87% of the students are likely to obtain this z score or a higher z score on this exam. In our new terminology, we would say that the probability (p) of obtaining this score or a higher score is 0.0287, so $p = 0.0287$.

Another Example—Calculating the Probability of a Particular z Score

Suppose the time from conception to birth in humans is approximately normally distributed, with a mean of 280.5 days and a standard deviation of 8.0 days. In a paternity case, it was demonstrated that the time from the alleged conception to the birth of a 7.0-lb (average size) baby was at least 299 days. One might argue that 299 days is significantly longer than you would expect by chance, and the baby might have been conceived later than the alleged conception date. You could evaluate the likelihood (or probability) of a gestation of 299 days:

Step 1: Compute the proportion of women having this or a longer gestation time, which you should now recognize as the probability that this pregnancy had a gestation length that came from a normally distributed population of gestation lengths.

Formula for a *z* Score

$$z = \frac{X - \mu}{\sigma}.$$

Applying the Formula to Our Example

$$z = \frac{299 - 280.5}{8} = 2.3125.$$

 Step 2: Evaluate the probability of obtaining this score due to chance.

From Table A, column C, the probability of obtaining a *z*-obtained score of 2.31 or greater from the normally distributed population is 0.0104.

How should we interpret these data in light of the paternity court case?

What would you tend to believe if you were the judge? These results suggest that about 1.04% of all normal pregnancy lengths are this long or longer. Thus, it is unlikely that the alleged conception date is accurate, and it is more likely that the conception took place later than was alleged. However, there is a small probability (1.04%) that this pregnancy did indeed take place on the proposed date. This is an important consideration when interpreting probabilities.

More Formally: Theoretical and Empirical Probability

> **Probability:** A number between 0 and 1 that indicates the likelihood of an event occurring due to chance factors alone.

Probability can be written as a mathematical expression of a specific event. For example,

p(A) reads as the probability of the occurrence of Event A or the probability of A.

> **Theoretical probability:** The number of events classifiable as A divided by the total number of possible events.

Classical, or **theoretical probability,** is the number of events classifiable as A divided by the total number of possible events. Theoretical probability is a priori, meaning that it is a probability that is formed on potential events, not actual events. It is the probability of rolling a dice and getting a 3, which is 1/6, since there are six sides to a dice. This is not the same as asking how many times a 3 came up when you actually rolled a dice 300 times. This expression is an empirical probability rather than a theoretical probability. **Empirical probability** is a probability that is calculated a posteriori (after the fact) and is the number of times A occurred divided by the total number of occurrences.

> **Empirical probability:** The number of times an event classifiable as A occurred divided by the total number of occurrences.

When you have a large sample size, the empirical probability approaches the theoretical probability. Thus, the actual probabilities will be most similar to the predicted probabilities when the sample size is large, and in fact, if your sample size is infinitely large, your empirical probability will be equal to your theoretical probability. This concept will have ramifications later for the effect of the sample size on the validity of your results. Specifically, we will be more confident of the results when our sample sizes are large.

Probability Notation

As we have shown, $p(A)$ is the notation used to suggest the probability of an event named A. Probabilities always range from 0 to 1.00, but it is acceptable to use fractions or decimals to express probabilities. For example, you can express probability as the "chances in one hundred." In scientific hypothesis testing, we typically accept an error rate that is equal to "five out of one hundred," which can also be expressed as a fraction, "5/100," or a decimal "0.05."

If you have been to the horse track, you are familiar with the terminology of odds. Odds statements are ratios that are given in favor or against the likelihood of an event. For example, you might bet on a horse where the odds show 3 to 1 in favor or 3 to 1 against. You can rephrase those odds to be 3 in favor and 1 against (same as 3 to 1 in favor) and 3 against and 1 in favor (same as 3 to 1 against). Thus, in the first case, ¾ are in favor, and in the second case, ¾ are against the likelihood of the event (a particular horse winning or placing in the race). Thus, in the second example, if ¾ are against, then ¼ or 25% are in favor of the event. The second case is considered "longer odds," or less likely. To easily convert odds statements into fractions, simply place the first number in the numerator (top) and then add the first and second numbers to get the denominator (bottom), and use the "in favor" or "against" wording to determine the direction of the predicted odds. For example, 8 to 1 against is 8/9 against. Thus, the odds are $8 \div 9 = 0.888888$ that the event will not occur, or $1.00 - 0.8888880 = 0.111111$ that the event will occur.

SUMMARY

In this chapter, you have learned about the normal distribution, standardized (z) scores, and probability. These concepts begin to build the infrastructure that you will need to learn inferential statistics. We will introduce more of these concepts in the next chapter. The definition of a normal distribution and the concepts of probability will be important throughout this book since many inferential tests require that your data be normally distributed so that the probability of obtaining a particular result by chance can be calculated. Thus, you will repeatedly calculate statistics on your data but interpret them based on the likelihood of obtaining that statistic simply due to chance, not because of your experimental manipulations. If the chance (or odds) is low, then you will interpret your results as unlikely due to chance or more likely as being the result of your experimental manipulation. We will continue this discussion in the next chapter.

FYI: CALCULATING DISCRETE PROBABILITIES

In this chapter, we have described the concept of calculating continuous probabilities, or the area under the curve and in the tails. This is a critical concept in statistics. Traditionally, statistics courses have also included a chapter on calculating discrete probabilities of events as a segue into discussion of continuous probabilities. Here we present the rules for calculating discrete probabilities for your review.

The easiest form of calculating discrete probabilities is the situation where you are calculating the probability (p) of only one possible event.

Example

Calculate the probability that you will randomly select a Pepsi from your fridge given that you have 5 Pepsis, 6 root beers, 3 orange sodas, and 8 grape sodas. We solve this problem by looking at the total number of sodas that are Pepsi and dividing them by the total number of canned sodas in the refrigerator.

$$p(\text{Pepsi}) = 5/(5 + 6 + 3 + 8) = 5/22 = 0.2273 \text{ or } 22.73\%.$$

Sometimes you want to determine the probability of one or another event occurring when there are multiple possible outcomes. In this situation, the calculation is slightly more complicated.

Addition Rule

If A and B are events, the probability (P) of obtaining either of them is the probability of A and the probability of B, or $P(A)$ and $P(B)$. So, if we want to calculate the probability of either A or B occurring, meaning that either outcome would be satisfactory, then we would add the probabilities together:

$$P(A \text{ or } B) = P(A) + P(B).$$

This rule or formula works well if either event will satisfy the conditions *and* it is not possible for A and B to occur simultaneously. In other words, Events A and B are mutually exclusive. An example of this would be the probability of giving birth to a boy or girl.

This formula is easily expanded if there are more than two mutually exclusive events:

$$P(A \text{ or } B \text{ or } C \text{ or } D \ldots \text{ or } Z) = P(A) + P(B) + P(C) + \ldots + P(Z).$$

Example

Pretend you have three sisters who are all currently pregnant. What is the probability that two or more of your sisters will give birth to baby girls?

Births	Sister 1	Sister 2	Sister 3	No. Ways This Is Possible	Odds	*p*
p(3 girls)	G	G	G	1	1/8	.125
p(2 girls)	G	G	B	3	3/8	.375
B	G	G				
G	B	G				
p(1 girl)	G	B	B	3	3/8	.375
B	G	B				
B	B	G				
p(0 girls)	B	B	B	1	1/8	.125
Total				8		1.00

$$P(2 \text{ or } 3 \text{ girls}) = P(2 \text{ girls}) + P(3 \text{ girls}) = .375 + .125 = .500 \text{ or } 50\%.$$

If A and B can occur simultaneously, then there is a modification of the formula to subtract the probability of both occurring simultaneously. Essentially, you are counting the chances of it occurring in A and in B, and you don't want to double-count the probability, so you subtract the chance of both occurring so you are not counting it twice. This modified formula is

$$P(A \text{ or } B) = P(A) + P(B) - P(A \text{ and } B).$$

Example

Suppose that you draw one card from a deck of 52 playing cards. Recall that in a deck of standard cards, there are 4 suits (clubs, spades, hearts, and diamonds) with 13 cards per suit (cards from 2 to 10

(Continued)

(Continued)

plus one ace, one king, one queen, and one jack). What is the probability that the card will be either a queen or a diamond? The probability of drawing a queen is 4/52, the probability of drawing a diamond is 13/52, and the probability of drawing the queen of diamonds is 1/52. So,

$$P(\text{Queen or Diamond}) = 4/52 + 13/52 - 1/52 = 16/52 \text{ or } 0.307692 \text{ or } 30.7692\%.$$

Multiplication Rule

So far, we have calculated discrete probabilities for when either Event A or B (etc.) can occur. Now let's turn to the case where both can occur and you actually want to know the probability of both occurring. In this case, you want to proceed as indicated in the rules above, but you want to multiply the probabilities rather than add them together. Note this will result in a smaller probability or chance/likelihood of the event occurring. This makes sense if one or the other won't satisfy the condition but you must have *both* occur, which can happen in fewer ways.

$$P(A \text{ and } B) = P(A) * P(B).$$

As before, you can expand the formula for more than two events or conditions. For independent events, the formula would be

$$P(A \text{ and } B \text{ and } C \text{ and } \ldots \text{ and } Z) = P(A) * P(B) * P(C) \ldots P(Z).$$

This is fairly easy if you understood the addition rule. It gets more complicated when you consider the effect of the order of events on the probability.

So, what is the probability of obtaining Events A and B given that the first event has already occurred (or the probability of B occurring given that A has already occurred)?

$$P(A \text{ and } B) = P(A) * P(B|A).$$

Example

Suppose that you are going to randomly order six individuals, A, B, C, D, E, and F. The probability the order will begin A B _ _ _ _ is

$p(A \text{ in first slot}) = 1/6$.

$p(B \text{ in second slot}) = 1/5$ (given that A has already occurred because you are sampling without replacement).

$p(A) * p(B) = 1/6 * 1/5 = .033$.

Therefore, the probability the order will have A in the first slot and B in the second is 3.3%.

The multiplication rule can be extended to more than two events with the following formula (this describes three events only):

$$P(A \text{ and } B \text{ and } C) = P(A) * P(B|A) * P(C|A \text{ and } B).$$

For more on the concept of sampling with and without replacement, see Chapter 7.

CHAPTER 5 HOMEWORK

Give short answers to these conceptual questions.

1. What is probability in statistics?

2. What does it mean to say that the odds are 5 to 1 against?

3. What does it mean to say that the odds are 2 to 1 in favor?

4. How far would you have to go away from the mean (on a normal curve) before the curve touched the axis (i.e., before the probability of a score was 0)?

5. What is the range of values for probability?

6. How is a priori probability different from a posteriori probability?

7. If you're looking at a normal curve, at what z scores would you find the inflection points?

8. Why can we compare sample data that have different means and standard deviations by using z scores?

Calculate the solutions for the following problems:

9. Find z scores for each data point in the following sample of test scores. Assume no inference is being made. Use the mean and standard deviation of the sample itself. 56 82 74 69 94

10a. Using a mean of 36.4 and a standard deviation of 2.7, find the score corresponding to a z score of 3.52.

10b. With this score and the above mean (36.4), did this subject do better or worse than a subject with the same score but where the standard deviation was 6?

Calculate the following for a normal distribution with a mean of 54 and a standard deviation of 3.2:

11. A z score for a raw score of 81:

12. A z score for a raw score of 51:

13. What raw scores bound the central 60% of the distribution?

For a nursing class having an average GPA of 3.1 and a standard deviation of 0.7. . .

14. What percentage of students have a GPA higher than 3.5?

15. What percentage have between 2.5 and 3.0?

Let's say that the mean for the first exam in another statistics class was 82 and the standard deviation was 4.8:

16. What is the z score for someone scoring 91?

17. What did a person score if his or her z score was –0.61?

18. For patients in a local clinic, the mean blood pressure was 130 mm Hg with a standard deviation of 16. The most extreme 5% of the scores fall beyond which blood pressure(s)?

19. A testing bureau reports that the mean for all students who took the GRE on October 31, 2001, was 500 with a standard deviation of 90. The scores are normally distributed. What is the raw score that lies at the 90th percentile?

PART II

INTRODUCTION TO HYPOTHESIS TESTING

CHAPTER 6

Sampling Distribution of the Mean and the Single-Sample *z* Statistic

Get ready to do your first "real" statistical assessment. You will now be able to statistically answer questions such as "Is this child's birth weight normal?" and "Are the murder rates of my city higher than the national average?" Before we get to those answers, we need to provide some conceptual groundwork so you know how we are able to answer these questions statistically.

In the previous chapter, you learned about the normal distribution, standardized (*z*) scores, and probability. The definition of a normal distribution and the concepts of probability will be important throughout this book since many inferential tests require that your data be normally distributed so that the probability of obtaining a particular result by chance can be calculated. Thus, you will repeatedly calculate statistics on your data but interpret them based on the likelihood of obtaining that statistic simply due to chance, not because of your experimental manipulations. If the chance (or odds) is low, then you will interpret your results as unlikely due to chance or more likely as being the result of your experimental manipulation.

The best way to summarize data from a sample is to look at the mean and standard deviation of a sample, and the best way to compare a sample and a population is to compare their means. Thus, instead of comparing one score to a population mean, we want to compare a sample mean to a population mean, and we want to evaluate the difference between the sample and the population means based on chance. How do we do that? Essentially, we need to evaluate our calculated statistic (the difference between the sample and the population means divided by the standard deviation of the population) with an appropriate sampling distribution.

In other words, there are two steps to answering a question about whether a particular sample came from the population.

1. You calculate the appropriate statistic.

2. You evaluate that statistic based on its sampling distribution.

> **Sampling distribution:** A hypothetical representation of all possible values of a statistic with the probability of each possible value if no independent variable is acting.

What is a **sampling distribution?** You have worked with one sampling distribution already, the normal distribution, and you used it to evaluate your z scores.

Most of this textbook involves teaching you a number of statistical tests, each with its own sampling distribution, and you will learn how to calculate the statistics as well as how to evaluate them based on chance, using the corresponding sampling distribution. In this chapter, we build on the concept of sampling distributions and probability, and this lays the foundation for the remaining chapters in this book.

HOW ARE SAMPLING DISTRIBUTIONS CREATED?

Three steps to creating a sampling distribution can be used to evaluate the statistics that you calculate (see Figures 6.1–6.4 for a graphical display of this concept).

1. Determine all possible samples of size n that can be drawn from the population.

2. Calculate the statistic that you are using for each sample.

Figure 6.1

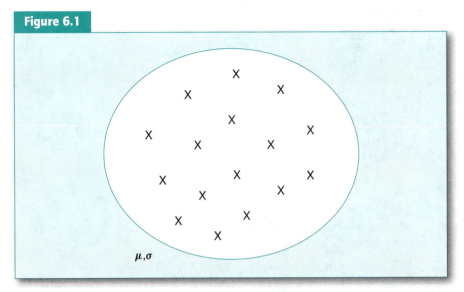

Sampling Distribution and Standard Error of the Mean: The Underlying Raw Score Population, With Parameters Mean and Standard Deviation

3. Calculate the probability of each statistical value based on chance alone. Essentially you are determining how often each statistical value occurs by chance.

By repeating this process for different sample sizes (n) and calculating the mean of each sample that you randomly draw from the **population,** you can create a **sampling distribution of the mean** that has these characteristics:

> **Population:** The entire set of scores.

> **Sampling distribution of the mean:** A theoretical distribution of means obtained by drawing all possible samples of a given size from a population and calculating a mean for each sample. From this distribution, the probability of obtaining any given mean from a population with the same parameters by chance can be calculated.

1. The mean of the sampling distribution of means (symbolized as $\mu \bar{X}$) is equal to the mean of the population of raw scores or $\mu \bar{X} = \mu$.

2. The standard deviation of the sampling distribution of means (symbolized as $\sigma \bar{X}$) is also called the standard error of the mean because each sample mean is an estimate of the mean of the raw score population. As an estimate, there will be some error associated with how close the estimate is to the true population values.

3. The standard deviation of the sampling distribution of means (the standard error) is equal to the standard deviation of the population of raw

Figure 6.2

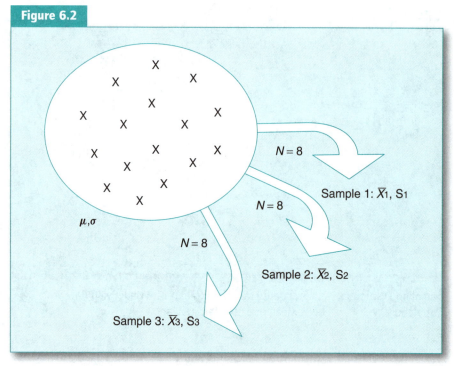

Repeated (Theoretical) Sampling of the Raw Score Population

scores divided by the square root of $n(\sqrt{n})$, so $\sigma\bar{X} = \frac{\sigma}{\sqrt{n}}$. As the sample size ($n$) increases, our estimate of the population mean improves. Thus, the standard error of the mean decreases with an increasing sample size. This makes sense because if we have a population of 1,000 scores, but we draw samples of $n = 2$, then our estimate of the mean based on each sample of $n = 2$ is not very good, and this would be reflected in our standard error. However, if we draw samples of 500, then our sample means will be close to the true population mean each time, and our standard error will be smaller.

4. The sampling distribution of the mean will be normally distributed depending on the shape of the population of raw scores and on the size of our sample. Having this sampling distribution be normally distributed is a very handy feature since we have seen how easy it is to work with probability and normal distributions. First, the sampling distribution of the mean will be normally distributed, regardless of the sample size used to create it, if the underlying raw scores are normally distributed. But if the raw scores are not normally distributed or if we don't know that they are, there is still hope.

Figure 6.3

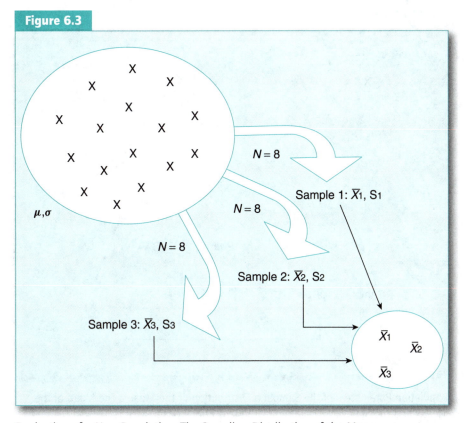

Production of a New Population: The Sampling Distribution of the Means

The central limit theorem (a mathematical proof of which you don't need the details: trust us!) tells us that regardless of the distribution of our raw scores, the sampling distribution of means will approach a normal distribution as our sample size increases. Typically, we assume that it will be normally distributed if our sample size is greater than or equal to 30 ($n \geq 30$). This assumption can be satisfied in a number of ways, including knowing that the population is normally distributed, showing that the sample is normally distributed or the sample size is > 30, or transforming data that are not normally distributed into normally distributed data by taking the log of each score. Taking the log of each score is only one way to transform data, and it doesn't always work. It is also important to note that while there is more than one way to determine if the sample is normally distributed, it does not mean that achieving normality is easy.

Figure 6.4

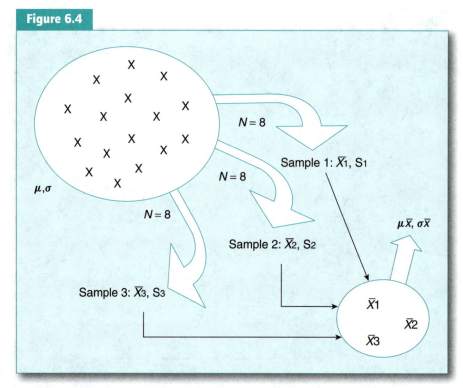

Population Parameters of the Sampling Distribution of the Means: The Mean of the Means and the Standard Deviation ("Standard Error") of the Means

Do you see how a normal distribution could have been created by calculating all possible values that z-obtained could take and then by assigning the likelihood of each value by how often it occurs? If not, review Figure 6.1 as a visual aid and imagine that we are taking each raw score and subtracting the population mean from it and then dividing by the population standard deviation. Imagine repeating that for every possible score. That would create a wide range of z-obtained values, some of which would be common (those would come from scores close to the mean), and others would be less likely (those would be far from the mean). We could assign probabilities based on how often each z score came about by random sampling of the raw scores. Figure 6.1 is demonstrating this same concept except that we are randomly drawing samples and then calculating the mean of the sample rather than calculating z scores. Nevertheless, the process is to randomly sample from the population, calculate a statistic, and determine the probability of that statistic occurring (whether it is a z score or a mean).

We are now ready to calculate the normal deviate z test, which is also called the single-sample z test.

SINGLE-SAMPLE z STATISTIC

In this section, we will convert the z score to a single-sample z test, which requires us to substitute some of the terms. Note the minor changes that we make to the z score formula.

Raw Score Formula

$$z = \frac{X - \mu}{\sigma}.$$

Mean Score Formula

$$z = \frac{\bar{X} - \mu_{\bar{X}}}{\sigma_{\bar{X}}}.$$

Mean Score Formula Simplified

$$z = \frac{\bar{X} - \mu}{\sigma_{\bar{X}}}.$$

Let's do a simple example with the numbers only—we'll do another example with a word problem later on.

 Step 1: Evaluate the values you are given.

Let's assume that you have drawn a sample from a population, and the sample mean $(\bar{X})$ is equal to 72 while you know that the population mean (μ) is equal to 75, and the standard deviation of the sampling distribution of means (or standard error) is equal to 1.66. Remember that the formula for the standard deviation of the sampling distribution of means ($\sigma_{\bar{X}}$) is equal to $\frac{\sigma}{\sqrt{n}}$, but we have given you $\sigma_{\bar{X}}$ in this example, so it is not necessary to divide by the square root of n this time.

Applying the formula to our example:

$$z = \frac{72 - 75}{1.66} = -1.88.$$

 Step 2: Evaluate the probability of obtaining this score due to chance.

From Table A, column C in the appendix, the probability of obtaining a z value of –1.88 (just look up the absolute value of 1.88 in the table) from the normally distributed population is 0.0301. We could conclude that it is unlikely that a sample with a mean of 72 comes from our null hypothesis population that has a mean of 75. We would conclude this based on the fact that 3.01% is a relatively low chance that this sample came from the original population of scores.

Another Example

Let's assume that you have drawn a sample of five ($n = 5$) sixth-grade children in the New Orleans school district from a population of schoolchildren in Louisiana. The mean number of correctly spelled words on a test with 17 words is recorded, and the sample mean ($\bar{X}$) is equal to 14 while you know that the population mean (μ) is equal to 15, and the standard deviation of the population is equal to 2.22.

★ **Step 1:** Evaluate the values you are given.

Remember that the formula for the standard deviation of the sampling distribution of means ($\sigma \bar{X}$) is equal to $\frac{\sigma}{\sqrt{n}}$. We have NOT given you $\sigma \bar{X}$ in this example, so it is necessary to divide by the square root of n this time.
Applying the formula to our example:

$$\sigma \bar{X} = \frac{\sigma}{\sqrt{n}}.$$

$$\sigma \bar{X} = \frac{2.22}{\sqrt{5.00}} = \frac{2.22}{2.2360679} = 0.9928142.$$

$$z = \frac{\bar{X} - \mu}{\sigma_{\bar{X}}}.$$

$$z = \frac{14 - 15}{0.9928142} = -1.007237 = -1.01.$$

★ **Step 2:** Evaluate the probability of obtaining this score due to chance.

From Table A, column C, the probability of obtaining a z value of –1.01 or lower (just look up the absolute value of 1.01 in the table) from the normally distributed population is 0.1562. We could conclude that it is likely that a sample with a mean of 14 comes from our population that has a mean of 15. We would conclude this based on the fact that 15.62% is a relatively good chance that this sample came from the original population of scores.

SUMMARY

In this chapter, we have extended the concepts of probability and z scores to consider the probability of obtaining particular sample statistics, such as means, from populations of means. Now we can ask questions about how likely it is that the sample mean comes from a particular population of sample means (the sampling distribution of means). The foundation for inferential statistics has now been laid, and we move on to introducing terminology for inferential statistics in the next chapter.

CHAPTER 6 HOMEWORK

Define the following terms.

1. What is the sampling distribution of a statistic?

2. What is the sampling distribution of the mean?

3. What is the central limit theorem?

Answer these short-answer questions.

4. Why is the standard deviation of the mean also called the standard error?

5. What are the three steps to creating a sampling distribution that can be used to evaluate the statistics that you calculate?

Perform these calculations and state your conclusion.

Questions 6 to 8 use the following information:
    The following table describes norms for performance of infants on developmental assessments (consider these to be population estimates). You may assume that these scores are normally distributed and that higher scores are "better." The sample size for these norms is 10,526. Results of the same assessments for a sample of 12 infants exposed prenatally to alcohol are also recorded:

Assessment	Norms		Fetal Alcohol Sample
	Mean	Standard Deviation	
a. Apgar	2.3	0.2	1.7
b. Recognition Memory	34	8.2	15
c. Object permanence	125	12.3	99

6. Calculate the experimental sample's z statistic for each assessment.

7. On which task did this sample do best? On which was the poorest performance? Explain how you arrived at this conclusion.

8. Individuals who fall in the lowest 10% of any assessment would be described as "at risk" for later developmental problems. For the Apgar task listed above, do the following: (a) calculate the "critical 10% score" below which an infant would be "at risk" and (b) decide whether you would consider the fetal alcohol infants to be at risk based on that assessment.

Questions 9 and 10 use the following information:

In a child development study, the age at which a cognitive developmental milestone is achieved is recorded for a large sample of unaffected children. The results are mean of 23.6 days, standard deviation of 2.3 days, normal distribution. A sample of 6 children who have been exposed to high levels of methanol (a toxic but common additive to gasoline) are brought into the clinic and tested. They reach the cognitive milestone at a mean of 28 days of age.

9. What is the probability that this sample of children was unaffected by methanol exposure and thus belongs to the statistical population of normal children?

10. Do you believe that this sample has been affected by methanol exposure (yes/no)?

11. If scores on an environmental health exam are normally distributed with a mean of 88 and a standard deviation of 9, what is the probability that your class of 100 students will score a mean of less than 90?

12. Rats enjoy running in wheels. Rats from a normal population will run an average of 133 revolutions in 15 minutes with a standard deviation of 15. If a sample of five rats that are exposed to caffeine run an average of 163 revolutions in 15 minutes, what is the probability that the caffeinated rat sample is drawn from the normal population?

Answer Questions 13 to 15 based on the following paragraph.

A random sample of 80 college students was selected to participate in an experimental anxiety reduction program designed to decrease test anxiety. After completing the anxiety reduction program, participants completed a battery of tests designed to measure anxiety level. The mean on the anxiety measures for the sample was 317 points. According to the researcher's vast data on anxiety, the mean anxiety level for college students who have not completed the anxiety reduction program is 310 points and a standard deviation of 48.

13. Compute the appropriate test statistic. Show your work for full credit.

14. Given these data, what would you conclude?

15. If the researcher did not have the standard deviation of scores for students who have not completed the anxiety reduction intervention program, would that have been the appropriate test statistic to compute?

CHAPTER 7
Inferential Statistics

In the last chapter, we discussed how you can statistically evaluate the difference between individuals and a population. Now we're moving ahead to lay out some details on how we can evaluate hypotheses using statistics. We will explain the logic used to ask whether there is a statistically significant difference between groups, such as a difference between the Apgar scores (method of measuring immediate health in newborns) of newborns born in New York versus New Guinea.

From this point on, you will be learning about inferential statistics. Now that you understand descriptive statistics, we can move you toward making inferences about populations based on sample data. First we define the relevant terms for this section, and then we introduce you to the logic of hypothesis testing critical values, power, effect size, and the distinction between parametric and nonparametric statistical tests. This chapter continues to lay the foundation for learning a large number of statistical tests, and Chapter 8 starts you on a review of the many tests we cover in this book.

Random sampling is a basic requirement for inferential statistics and is necessary so that one can infer that the sample is representative of the population, and thus the population characteristics can be inferred from the sample by applying the laws of probability. Random sampling occurs when subjects are selected in a manner that ensures that each possible sample of a given size is equally likely to be selected and each subject in the population is equally likely to be selected for the sample.

> Random sampling: Subjects are selected in a manner that ensures that each possible sample of a given size is equally likely to be selected and each subject in the population is equally likely to be selected for the sample.

HOW ARE SAMPLES RANDOMLY SELECTED?

Researchers often use computers or random number tables to select samples. Random number tables are effective if you pick your first number randomly, for example, by closing your eyes and pointing to a number as a way to enter the table and begin.

Using a computer to select a random sample can be problematic since computers are not random but are programmed to select numbers in a particular way, so that samples drawn with the use of a computer are pseudo-random rather than truly random. Most personal computers now use complex algorithms to manipulate an arbitrary starting value sufficiently to produce a resulting number that, for all intents and purposes, appears random.

Samples can be randomly selected in one of two ways. **Sampling with replacement** is performed such that each subject that is selected for a sample is returned to the pool prior to the next selection. **Sampling without replacement** is performed when each selected subject is not returned to the subject pool prior to the next selection. There is an interesting paradox that develops with these concepts. Sampling with replacement is a mathematical assumption for almost all inference tests. However, for practical purposes, most researchers use sampling without replacement to select subjects. In fact, many researchers think that it is better for them not to use a subject more than once in an experiment, which may be true if the measured variable might be affected by subject awareness of experimental procedures, for instance. How do we get out of this paradox? It turns out that if the sample is very small relative to the size of the population, then the effect of sampling without replacement

> Sampling with replacement: Each subject that is selected for a sample is returned to the pool prior to the next selection.

> Sampling without replacement: Each selected subject does not return to the pool prior to the next selection.

on the mathematical assumptions is minor. In fact, this assumption is routinely ignored by most researchers, but violation of this assumption (by having a large sample relative to the population size) can have serious effects on results, leading to erroneous conclusions on the relationship between the tested variables.

HYPOTHESIS TESTING

Researchers usually conduct an experiment because they have a hypothesis that an independent variable affects a dependent variable (**alternative hypothesis**) or that there is some relationship between the variables. They wish to contrast this with the possibility that the two variables are unrelated or that any patterns that are seen are due to chance factors rather than their experiment (**null hypothesis**). These effects can be directional (predicted to increase or to decrease) or nondirectional (the direction of the effect is not stated). **Nondirectional hypotheses** are more common since the researcher must have a prior scientific basis for predicting a specific direction to the effect. In fact, predicting a directional effect when the actual effect actually turns out to be in the opposite direction means that the experiment must be repeated. So while most researchers really have a **directional hypothesis** in mind, they actually test a nondirectional hypothesis to be conservative.

> **Alternative hypothesis (H_A):** There is an effect or relationship between two variables.

> **Null hypothesis (H_0):** Any effect or relationship between two variables is due to chance factors.

> **Nondirectional hypothesis (two-tailed):** The direction of the effect is not indicated in the alternative hypothesis, and the null hypothesis suggests only that there is no effect or relationship between the variables.

> **Directional hypothesis (one-tailed):** The direction of the effect is indicated in the alternative hypothesis, and the null hypothesis suggests the opposite or no effect.

> **Real effect:** When your independent variable affects your dependent variable.

The statistical process that we have been describing asks us to make an inference about **real effects.** Real effects are the actual effects (perhaps lack thereof) that the independent variable has on the dependent variable, out there in the real world. Researchers collect data to estimate the real effect of their independent or predictor variable, but we can never truly know the real effect of a variable. Instead, we can only make inferences about this effect. This inference is measured as the relationship between two variables or the effect of one variable on another, but since we do not know what the real effect is, it would be circular logic to try to say that what we see is the real effect. Instead, we take a different approach: Can we say that what we have recorded is *not* likely due to simple chance effects? An explanation based on simple chance effects is our null hypothesis. So we can only use statistics to test whether the null hypothesis is false.

We know how to deal with chance effects through the principles of probability: We have done so in the case of single scores and the probability of getting a particular score from a population with a known mean and standard deviation. So we can assign a probability to an explanation that our results are simply due to chance. If we fail to reject that explanation, then we must suggest that there is no real effect. If we reject that null hypothesis explanation, then we must turn to the only other explanation (in a clean experiment where all other factors have been controlled) and suggest that there is a real effect.

Rejecting the null hypothesis because there is overwhelming information suggesting that it is false still *does not prove* that the alternative hypothesis is true. If we are not able to reject the null hypothesis, we still cannot *accept* it, but we only *fail to reject* the null hypothesis. Thus, we can never state beyond a doubt that we have proven anything! This may make the scientific process seem pointless, but recall that part of the scientific method is in repeating experiments and testing alternative hypotheses. With this process, scientists are able to get closer and closer to the underlying truth. Nevertheless, because we rely on the laws of probability, and there is always a small probability that our results are due to chance factors alone, we cannot prove effects beyond all doubt. However, when specific effects are found repeatedly, and the likelihood that the effects are due to chance are small each time the experiment is conducted, the plausibility of a real effect of our independent variable becomes more and more accepted over time.

Critical Level of Probability

To test hypotheses, researchers calculate a statistic from their data, which, in some way, describes the magnitude of the effect of the independent variable, and then ask how likely it would be to obtain that statistic if chance factors alone were at work. If obtaining that statistic is very unlikely under chance conditions (less than or equal to 0.05 or 5%), then the researcher can reject the null hypothesis (which states that there is no relationship between their two variables). Thus, the researcher is suggesting that if their outcome would only occur in 5% or fewer cases under normal circumstances, then the likelihood that it occurred due to chance is low, and it is more likely that the variables in their experiment affected the outcome. Researchers must choose the critical level of probability for their experiment.

In this case, we have chosen 0.05 or 5% because that is the most commonly accepted critical level of probability in science. The critical level of probability is called the alpha (α) level. While 0.05 is the most commonly used alpha (α) level, 0.01 (1%) and 0.10 (10%) are used fairly often as well. We will discuss the pros and cons of each level in the next section.

Types of Errors

Our statistics are intended to allow us to assign a probability to whether there is a real effect in our experimental design. By choosing a critical probability and comparing our calculated null hypothesis probability to that critical value, we can make a statistical decision about the existence of a real effect: There is a real effect (reject the null hypothesis) or there is not a real effect (fail to reject the null hypothesis). In making this decision (i.e., in calculating our null hypothesis probability), it is possible for us to make an error.

Researchers and statisticians can make two types of error when using statistics to evaluate hypotheses. **Type I error** is when you reject the null hypothesis when it is actually true (you conclude that there was an effect when there was not), and **Type II error** is when you fail to reject the null hypothesis when it is actually false (concluding that there was no effect when there was). It sometimes helps to keep track of these mentally if you think of Type I error as rejecting the null when you shouldn't and Type II error as failing to reject the null when you should reject it. If you reject the null hypothesis when in reality it is false or you fail to reject the null hypothesis when in fact it is true, then you have made correct decisions (see Table 7.1).

Type I error: Error that occurs when you reject the null hypothesis when it is actually true.

Type II error: Error that occurs when you fail to reject the null hypothesis when it is actually false.

Table 7.1

Statistical Decision	Null Hypothesis True (No Real Effect)	Null Hypothesis False (Real Effect)
Fail to reject null hypothesis	Correct decision	Type II error
Reject null hypothesis	Type I error	Correct decision

Reality

Clearly, it is desirable to limit the likelihood of either type of error. Alpha (α) is the limit we set on Type I error, and beta (β) is the limit we place on Type II error. Recall that the critical level of probability (α) is often set at 0.05. That means that we often choose to accept a Type I error rate of 0.05 or less. While it might seem desirable to limit that error rate even further, the lower you make α, the higher you make the likelihood of making the opposite type of error (failing to reject the null hypothesis when it is in fact false). Thus, there is a relationship between α and β such that as α increases, β decreases. Alternatively, as α decreases, β increases. Fortunately, statisticians have found that an $\alpha = 0.05$ generally gives an acceptable β level. Let's do an example to demonstrate the concepts covered so far in this chapter.

Example

A researcher wants to know if people can solve word puzzles faster when they have ingested caffeine. He designs a study where half of a group of students first solve a word puzzle under normal circumstances. Then the same students drink several ounces of caffeinated coffee and solve another word puzzle. The other half of the group solves the puzzle under the caffeine condition first and later solves another word puzzle when they are not under the influence of caffeine. He records the time it takes for people to solve the puzzles in both conditions. We choose to set $\alpha = 0.05$.

1. What are the researcher's null and alternative hypotheses?

2. Suppose that the researcher runs the experiment and collects and analyzes the data. He gets a p value of .042, and the times were shorter with caffeine. What does this p value mean, and how should the researcher evaluate the null hypothesis?

 Step 1: Determine the H_0 and H_A.

The researcher believes that caffeine will increase the speed of solving a word puzzle. Thus, the alternative hypothesis is directional.

H_0: There is no effect of caffeine on solving a word puzzle *or* caffeine decreases the speed of puzzle solving.

H_A: Caffeine increases the speed of solving a word puzzle.

Step 2: Evaluate the H_0.

> *Decision rule*
>
> > If $p < \alpha$, then reject H_0
> >
> > If $p > \alpha$, then fail to reject H_0
>
> *Decision*
>
> > $p = .042$
> >
> > $.042 < 0.05$, so we reject H_0

Conclude that caffeine appears to increase the speed of solving a word puzzle, but there is a chance (= .042 or 4.2%) that we have made a Type I error and falsely rejected the null hypothesis—that, in fact, there is *no* effect of caffeine on the speed of solving a word puzzle.

ISSUES FOR DEBATE: SHOULD WE EVEN USE HYPOTHESIS TESTING?

Recently there has been controversy about whether hypothesis testing is even appropriate. A number of psychologists and statisticians are arguing that other techniques are more appropriate, including the necessity of reporting means and standard deviations between groups/conditions, a greater reliance on graphical display of the data in publications, and reporting the estimated sample means for each group ± a measure of error. This last option is called a confidence interval, which we will discuss in a later chapter.

The controversy has led to the formation of a task force by the American Psychological Association (APA). The task force produced an article wherein they advocate choosing a middle ground between banning hypothesis tests altogether and maintaining the status quo. For more information on this topic, see Harris (1997) and Loftus (1991, 1993, 1996).

POWER

> Power: A measure (between 0.00 and 1.00) of the ability of a statistical test to detect an effect of the independent variable if the effect exists in reality.

Power is a measure of the ability of a statistical test to detect a real effect of the independent variable if the effect exists in reality. It is the probability of (correctly) rejecting the null hypothesis if it is false. The values of power range from 0.00 to 1.00, since it is in actuality a probability. Obviously, it is desirable to have the highest possible power in your statistical work and thus the highest possible probability of making a correct decision about the existence of the real effect.

A number of factors influence the power of your statistical comparison, and thus power is important from the moment you design your experiment. The issue of the power of your comparison will also come up if you fail to reject your null hypothesis. As we saw in Table 7.1, there are two possible reasons that you may fail to reject your null hypothesis: (1) You have made a correct decision, or (2) you have made a Type II error (β). The lower the power of your comparison, the more likely you will make a Type II error (β) because

$$\text{Power} + \beta = 1.00$$

or

$$p(\text{rejecting H}_0 \text{ if false}) + p(\text{failing to reject H}_0 \text{ if false}) = 1.00.$$

Values of power from .4 to .7 may be acceptable, but .8 or higher is considered good. Thus, if you fail to reject your null hypothesis but you calculate the power of your test and determine that the power was .28, you have good reason to suspect that your power was too low to detect a potential real effect of your independent variable, rather than the quick decision that there was no real effect.

What Affects Power?

1. *Magnitude of the real effect of your independent variable*

   As the size of the real effect increases, power increases ($\uparrow$ real effect = $\uparrow$ power). This is, of course, very hard to manipulate, but in some cases, a researcher could refine the size of the effect that they can expect to detect with their data.

2. *Sample size*

   As the size of your sample increases, power increases ($\uparrow n = \uparrow$ power).

3. *Alpha (α) level*

   Lowering α (which controls your chances of rejecting the null hypothesis when it is actually true) reduces your power, but recall that increasing α will also increase your Type I error rate ($\downarrow \alpha = \downarrow$ power and $\uparrow \beta$, but $\uparrow \alpha = \uparrow$ Type I error).

4. *Experimental design*

Experimental designs that control for the effect of confounding variables increase power by reducing sources of unexplained variability. We will describe some of these experimental designs later, but remember the point that reducing sources of variability in your data will increase power.

5. *Statistical test*

Statistical tests that require inferences about the population (parametric tests) are more powerful than statistical tests that do not require inferences (nonparametric tests). See the following section for more specifics on parametric versus nonparametric tests (parametric tests = ↑ power).

6. *Type of hypothesis*

Directional tests are more powerful than nondirectional tests (directional hypotheses = ↑ power) if the effect is in the hypothesized direction.

Conceptual Example

Assume that eating a candy bar raises your intelligence quotient (IQ) score by one point. Thus, chocolate has a small real effect on your IQ. IQ is measured by a standardized test where the $\mu = 100$ and $\sigma = 16$. What if you randomly sample two groups of five individuals from the United States? You give the first group (control group) a bagel to eat, and then you give them an IQ test. You give the second group (experimental group) a candy bar to eat, and then you give them an IQ test. You get the following results:

Control Group	Experimental Group
$\bar{X} = 100$	$\bar{X} = 101$
$n = 5$	$n = 5$

Notice that the mean IQ is one point higher in the experimental group compared to the control group. But how certain are you that it is because of the candy bar and not just a random effect of whom you assigned to each group? What is the probability of getting a sample with a mean of 101 or higher when your population mean = 100? You'll learn how to calculate that in the next chapter using a modified z score formula, but we did the work for you this time and found that $p = .4761$. That means that 47.61% of the time, you could get a sample mean of 101 or higher simply due to chance. How certain would you be that the effect was due to the candy bar if you didn't know ahead of time that candy bars increase IQ by one point? You would not be very certain, particularly since candy bars have such a small real effect on IQ score. If candy bars had the effect of increasing IQ score by 50 points (to 150 points instead of only 101 points), then we would be more certain that the candy bar was the reason for the increased sample mean in the experimental group because the difference between the two means would be so much larger. Likewise, if you sampled 200 people in the control group and 200 people in the experimental group, instead of only 5 in each group, we would be much more likely to believe even small differences between the two sample means could be real effects.

How Do You Calculate Power?

To calculate power, you need to consider the factors that influence power (magnitude of the effect of the independent variable, sample size, alpha level, experimental design, statistical analysis, and whether you are testing a one-tailed or two-tailed hypothesis). In the past, most people relied on published tables and power curves to estimate power or determine an adequate sample size, but it is now more common to calculate power since a number of software packages (e.g., SamplePower 2.0 by SPSS) have been developed to make the process simple. One good website to use for power calculations is UCLA's Department of Statistics website, which has a nice power calculator for multiple statistical analyses (www.stat.ucla.edu/).

Demonstration of Power

Here we provide a simple demonstration of how power is affected by sample size, the size of the real effect, and alpha (α). Here we have calculated power for you, and you can observe the influence each of these factors has on a few sample experimental situations. Recall that power calculations between .4 and .7 are considered acceptable by many researchers, but .8 or higher is considered a good power level for an experiment. Note that the actual power calculation values in this demonstration apply only for one statistical test (binomial test that we will not explain here).

Calculation of Power Example

Sample Size (n)	Size of the Real Effect	Alpha (α)	Power Calculation
4	All levels	0.05	**0.0000**
12	0.70 (moderate effect size)	0.05	**0.2528**
12	0.90 (large effect size)	0.05	**0.8891**
25	0.70 (moderate effect size)	0.05	**0.5118**
25	0.70 (moderate effect size)	0.01	**0.3407**
50	**0.70 (moderate effect size)**	**0.05**	**0.8594**

The table demonstrates that when your sample size is 4 and $\alpha = 0.05$, there is no power in your experiment, regardless of the size of your real effect. In this situation, you would *never* be able to reject the null hypothesis, no matter how large the real effect! However, if you increase your sample size to 12 and you have a moderate effect of your independent variable, you get a power result that is not zero but is still too low. The only way to get a larger power with a sample size of 12 is to have a real effect that is larger than .70, such as .90. Unfortunately, the size of your real effect is not always under the researcher's control. It is much more likely that the size of the sample is under your control, and thus we can increase the sample size to 25 with the same moderate real effect (.70) that we used with the smaller size and find that our power is now approaching the acceptable range (0.5118). However, note the effect on the previous power calculation when you reduce your alpha level to 0.01: Your power is diminished to 0.3407.

FYI: THE PROBLEM OF TOO MUCH POWER

Believe it or not, it is possible to have too much power. Specifically, that can occur when the researcher has a large sample size. If your sample is too large, nearly any difference will be "statistically significant." This can result in findings that are statistically significant but psychologically or biologically irrelevant. Besides leading to potentially erroneous results, having a very large sample size may waste time, resources, and money. Clearly, estimating power prior to collecting data can help researchers determine what sample size is large enough to be representative of the population but not so large that psychologically irrelevant differences are statistically significant. For more information on the issue of sample size, see Cranor (1990).

EFFECT SIZE

The concept of effect size is important for interpreting the importance of your results, and its calculation is required by many journals that publish scientific results and by many granting agencies prior to funding.

Effect size is a measure or estimate of the strength of the relationship between two variables. It can be calculated for a population or estimated for a sample.

The calculation of effect size is accomplished in slightly different ways with each statistical test and will be covered in the relevant future chapters. However, in general, it is a measure of the size of the difference between your groups (independent variables) weighted by the variability in each group (standard error). For example, if you knew that there was a large difference between the groups or conditions, it would also be important to know if most subjects within that group experienced that or if there is a wide variability in subject scores within each group. What if your groups were testing the effectiveness of alternative treatments for cancer? It's important to know not only the differences between the treatments but also how much variability there was in patient response within each drug treatment before we could understand how well the drug worked.

PARAMETRIC VERSUS NONPARAMETRIC TESTS

Parametric statistical tests require assumptions about the parameters of the population while nonparametric tests do not require assumptions about the population. While all of the statistical tests you will encounter in this book require random sampling, only parametric tests require that the underlying population is normally distributed. This assumption can be satisfied in a number of ways, including simply knowing that the population is normally distributed, showing that the sample is normally distributed or the sample size is > 30, or transforming data that are not normally distributed into normally distributed data. Despite the multiple ways to potentially satisfy these assumptions, it is not necessarily easy to meet the assumptions, and researchers should be careful to check the assumptions required for their analyses. It is also important to note that the sample size of 30 is a rule of thumb, and even that doesn't work if the sample data are very skewed.

Transforming raw data into normally distributed data can sometimes be accomplished by taking the logarithm, square root, and multiplicative inverse of each score. Because you transform each and every score in your sample, you maintain their relative differences and can then proceed with your analyses of these transformed scores. Of course, the actual data points are easier to interpret, so you typically graph your raw scores, but you perform parametric data analyses on the transformed scores. However, transformation does not always result in making your data normally distributed. If transformation does not work, then you must turn to a less powerful nonparametric statistical test for your analyses. There are some exceptions to this, including resampling tests (see Chapter 15 for a discussion of this topic).

> **Parametric tests:** Statistical tests that require assumptions about the parameters of the population.

> **Nonparametric tests:** Statistical tests that do not require assumptions about the underlying parameters of the population.

When all other factors that influence power are held constant, **parametric tests** are more powerful than **nonparametric tests.** This is because you gain a mathematical advantage if you can assume you know something about the parameters of the population. Of course, if you assume that incorrectly, your statistical decisions may be inaccurate. In the next chapter, you will learn how to calculate and evaluate your first parametric tests (single-sample z and single-sample t tests). You will find that these tests require knowledge of some parameters of the population (both require knowledge of μ and one requires knowledge of σ) and an underlying normal distribution.

SUMMARY

This chapter has addressed the process that one uses to evaluate whether a particular experimental result is due to the experimental manipulation or due to chance. This process is dependent on random sampling, but random sampling is not a guarantee that we will always make the correct decision about the meaning of our experimental results. Thus, there is always the chance that we have unknowingly made a Type I or Type II error, but most of the time, our experiment and our analysis should be done carefully enough that we should make correct decisions.

In any statistical analysis, there are three important values: the power, or sensitivity of your design to detect a real effect; the probability value, or probability that your results are due to chance; and the effect size, the magnitude of the effect. These are three different values, all related but important for the interpretation of any results.

In the next chapter, we will return to the single-sample z test, but now we'll consider it a hypothesis test using the terminology and procedures reviewed in this chapter. We'll also introduce a second single-sample test, the single-sample t test.

CHAPTER 7 HOMEWORK

Provide a short answer for the following questions.

1. What is a "real effect"?

2. What happens to power as your alpha level changes?

3. Describe the paradox produced by the difference between sampling with replacement and sampling without replacement.

4. Why do all statistical inferences require random sampling?

5. Describe a random sample.

6. What effect does increasing your sample size have on power?

7. What are some factors that influence the power of an experiment?

8. You decide that an alpha level of .05 is too high and want to reject the null hypothesis only if the probability is 1% or less than your results due to chance. How does your decision affect the probability of making a Type II error?

9. You are testing an herbal memory aid that has already been shown to have few side effects. You can only release this drug if the Food and Drug Administration gives you its approval. Given this situation, do you think it would be justified to set alpha at something other than .05?

10. What is the difference between a parametric and a nonparametric test?

Answer the following multiple choice questions.

11. With alpha and the effect of the independent variable held constant, as N increases, _____.

    A. power increases

    B. the probability of a Type I error increases

    C. the probability of a Type II error increases

    D. power decreases

12. With the effect of the independent variable and N held constant, as alpha becomes more stringent, _____.

    A. power increases

    B. power stays the same

    C. power decreases

    D. beta decreases

13. If H_0 is true and the probability of making a Type I error is .05, then the probability of making a correct decision is _____.

   A. power

   B. .05

   C. .95

   D. Need more information

14. Maximizing the power of an experiment _____.

   A. minimizes alpha

   B. minimizes beta

   C. increases the probability of rejecting H_0 when H_0 is true

   D. increases the probability of making a Type II error

15. With other factors held constant, as the effect of the independent variable decreases, power will _____ and the probability of a Type II error will _____.

   A. decrease, increase

   B. decrease, decrease

   C. increase, decrease

   D. increase, increase

16. The power of an experiment is affected by _____.

   A. the alpha level

   B. the sample size

   C. the magnitude of the independent variable's effect

   D. All of the above

   E. A and B

   F. A and C

Answer Questions 17 to 21 using the following information. A researcher is interested in whether marital therapy affects the happiness of each member of the couple. Twelve couples from New York City are randomly sampled for an experiment. Each member of the couple is given an assessment where they rate their marital satisfaction and their overall happiness prior to 6 months of marital therapy. After the therapy sessions are completed, each individual fills out the assessments a second time. A statistician conducts an appropriate statistical analysis to calculate the probability that the improvement they found after 6 months of therapy could be due to chance factors rather than due to the therapy intervention.

17. What is the nondirectional alternative hypothesis?

18. What is the null hypothesis that is appropriate for the nondirectional alternative hypothesis?

19. The obtained probability for this experiment is $p = .0384$. Using an $\alpha = .05$ (two-tailed), what should you conclude?

20. What is the population to which these results apply?

21. What error might you be making based on your conclusion in Question 19?

Answer Questions 22 to 24 using the following information. Pretend you're in graduate school and have just finished running a study looking at performance following negative verbal feedback. Your study used the same participants for both conditions (before and after feedback). Now you want to look at whether negative verbal feedback had any effect on participants' test performance.

22. What is the alternative hypothesis?

23. What is the null hypothesis?

24. If you calculate a p value of .0118, did negative verbal feedback have any effect on test performance? Use an alpha level of .05.

Answer Questions 25 to 27 using the following information. You believe students taking exams do better when the exams are taken in the afternoon rather than in the morning. To test your idea, you pair students in two sections of a psychology course, one in the morning and the other in the afternoon (the section instructor, GPA, and other extraneous factors are controlled for).

25. What is the alternative hypothesis?

26. What is the null hypothesis?

27. If your means are $\bar{X}_{morning} = 75.8$ and $\bar{X}_{afternoon} = 72.3$, and you calculate a p value of .0328, what do you conclude using an alpha level of .05?

CHAPTER 8

Single-Sample Tests

In this chapter, you will begin calculating your first hypothesis tests. This chapter will combine the concepts you learned in Chapter 6 on sampling distributions and probability with the concepts of hypothesis testing from Chapter 7. In Chapter 6, you acquired the skills to compare a sample mean to a population that was normally distributed and to interpret that sample mean based on the area under the curve associated with that score or the probability of obtaining that sample mean based purely on chance. In this chapter, you will test a hypothesis about whether a particular sample mean belonged to a population of normally distributed scores with only a small extension of these concepts.

We start off by calculating a *z* test, and then we present two complete story problems for *z* tests. We walk the student through the calculations, the interpretations, and then how you would represent this statistical test in a professional publication. Next you are introduced to the single-sample *t* test, again with two complete story problems. Finally, we introduce the concepts of choosing the appropriate statistical test, confidence intervals, power, and effect size. These last concepts will be covered in more detail in subsequent chapters.

SINGLE-SAMPLE z TEST

Recall the formula for the single-sample z statistic presented in Chapter 6:

Mean Score Formula

$$z = \frac{\bar{X} - \mu_{\bar{X}}}{\sigma_{\bar{X}}}.$$

Mean Score Formula Simplified

$$z = \frac{\bar{X} - \mu}{\sigma_{\bar{X}}}.$$

Let's do a simple example with the numbers only—we'll do another example with a word problem later on.

 Step 1: Evaluate the values you are given.

Let's assume that you have drawn a sample from a population, and the sample mean ($\bar{X}$) is equal to 72, the population mean (μ) is equal to 75, and the standard deviation of the sampling distribution of means (or standard error) is equal to 1.66. Remember that the formula for the standard deviation of the sampling distribution of means ($\sigma_{\bar{X}}$) is equal to $\frac{\sigma}{\sqrt{n}}$, but we have given you $\sigma_{\bar{X}}$ in this example, so it is not necessary to divide by the square root of n this time.

Applying the formula to our example:

$$z = \frac{72 - 75}{1.66} = -1.88.$$

 Step 2: Evaluate the probability of obtaining this score due to chance.

Evaluate the z-obtained value based on alpha (α) = 0.05 and a one-tailed hypothesis.

From Table A, column c (see the Appendix) the probability of obtaining a z value of −1.88 (just look up the absolute value of 1.88 in the table) from the normally distributed null hypothesis population is 0.0301, and 0.0301 < 0.05.

Our obtained probability of 0.0301 is less than our α of 0.05, and thus we reject the null hypothesis that suggests there is no difference between our sample and the population or that the difference is in the opposite direction that was outlined in the alternative hypothesis. We conclude that it is unlikely that a sample with a mean of 72 comes from our null hypothesis population that has a mean of 75.

A Complete Example

The Washington National Primate Research Center has maintained birth weight records for 40 years on 600 newborn pigtailed macaque monkeys. The birth weights are normally distributed, with a mean of 425 grams and a standard deviation of 60 grams. In the 1990s, they were trying to determine the effects of azidothymidine (AZT) on fetal development. AZT is a drug used to treat human immunodeficiency virus (HIV) infection, but at that time it was unknown whether HIV-positive pregnant mothers could take the drug without harming their unborn children. Four pregnant pigtailed macaques were given AZT during pregnancy. None of the monkeys was infected with HIV, so the study was to determine the effects of AZT alone on fetal development. One of the measures used to assess the effect of AZT on fetal development was birth weight, and those scores are listed below.

Subject	Birth Weight (g)
1	340
2	410
3	365
4	315

Null hypothesis: There is no difference between the mean birth weight of a sample of AZT-exposed pigtailed macaques and mean of the null hypothesis (untreated) population.

Alternative hypothesis: There is a difference between the mean birth weight of AZT-exposed pigtailed macaques and the null hypothesis population.

 Step 1: Compute the probability of the mean birth weight of this sample given that the sample comes from the null hypothesis population of normally distributed birth weights.

Formula for a z test:

$$z = \frac{\bar{X} - \mu}{\sigma_{\bar{X}}}.$$

Substitution of $\sigma_{\bar{X}}$:

$$\sigma_{\bar{X}} = \frac{\sigma}{\sqrt{n}}.$$

Calculate the mean of the sample:

$$\bar{X} = \frac{(340 + 410 + 365 + 315)}{4} = 357.5 \text{ grams}.$$

Applying the formula to our example:

$$z = \frac{357.5 - 425}{\dfrac{60}{\sqrt{4}}} = \frac{-67.5}{30} = -2.25.$$

 Step 2: Evaluate the probability of obtaining this score due to chance.

Evaluate the z-obtained value based on alpha (α) = 0.05 and a nondirectional hypothesis. From Table A, column c (in the Appendix): the probability of obtaining a z-obtained score of –2.25 (look up the absolute value of 2.25 in the table) from the normally distributed null hypothesis population is 0.0122. However, the hypothesis we are evaluating in this case is nondirectional or two-tailed, and the table only gives you the area under the curve (or probability) for one tail. Thus, we need to double that probability.

Double the probability for two-tailed evaluation:

$$0.0122 \times 2 = 0.0244.$$

Compare the two-tailed probability with our alpha level:

$$0.0244 < 0.05.$$

Our obtained probability of 0.0244 is less than our α of 0.05, and thus we reject the null hypothesis that suggests there is no difference in birth weight.

Alternatively, you could compare a z-critical value with your z-obtained value. When α = 0.05 and your hypothesis is two tailed, you should use the z-critical value that lists 0.025 in column c (doubled, that would be 0.05). Thus, the z-critical values for this example would be ±1.96.

To reject the null hypothesis, the absolute value of z obtained must be equal to or more extreme than (greater than) the z critical.

$$|Z_{\text{obtained}}| \geq |Z_{\text{critical}}|$$

$|-2.25| \geq |1.96|$, so we reject the null hypothesis (see Figure 8.1 on next page).

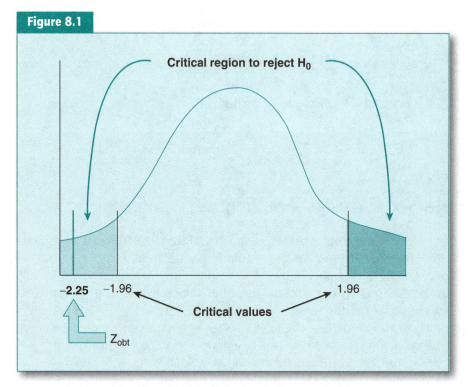

Figure 8.1

Here the z-critical values as well as the z-obtained values are shown on the X-axis. Clearly, the z-obtained value of −2.25 falls in the critical region to reject the null hypothesis. Drawing your own sketch of the z distribution is a helpful aid to making the correct decision about rejecting or failing to reject the null hypothesis.

How should we interpret these data in light of the risks of AZT on fetal development?

Based on a sample size of 4, these results suggest that only about 2.44% of all samples of normal pregnancy birth weights would have a mean birth weight this extreme or more extreme (farther from the mean, in either greater or lesser direction because we used a nondirectional hypothesis). Thus, it is unlikely that this sample of birth weights comes from the normal null hypothesis population of birth weights, and it is more likely that AZT affects birth weight. However, there is a small probability (2.44%) that these are normal birth weights that were not affected by AZT.

How would these results be reported in a scientific journal article?

For a journal that required American Psychological Association (APA) format, the results could be reported in a format such as this:

There was a significant difference between the mean birth weight of AZT-exposed pigtailed macaques ($M = 357.5$) and the null hypothesis population ($\mu = 425$), $z(N = 4) = -2.25$, $p = .02$.

This formal sentence includes the dependent variable (birth weight), the independent variable (AZT exposed or non–AZT exposed), as well as a statement about whether the results are (or are not) significant in a statistical sense, the direction of the effect indicated by reporting of the means, the symbol of the test (z), the sample size, the statistical value (-2.25), and the exact probability of obtaining this result simply due to chance ($.02$). Note that journals sometimes require that the mean is expressed as M rather than $\bar{X}$ or μ.

Keep in mind that the specific journal (e.g., *Child Development, Journal of Personality and Social Psychology*) may not strictly adhere to the APA standards due to editorial convenience. Journals such as *Animal Behaviour* or the *British Journal of Social Work* may deviate more dramatically from APA format.

A Second Complete Example

An industrial psychologist has been hired to advise on the establishment of a new manufacturing facility, a facility that must exhibit maximum efficiency of its workers to be competitive in the marketplace. To establish some baseline information, the psychologist measures job efficiency for four manufacturing lines, a measure for which there are considerable established data from other manufacturing facilities. In fact, it has been established that, industry-wide, efficiency has a mean of 2.8 widgets per day per line, with a standard deviation of 0.3 widgets per day per line. He collects the following data on widget production in the new plant:

Manufacturing Line	Widget Production, per day
1	2.8
2	3.2
3	3.4
4	2.6

Null hypothesis: There is no difference between the mean efficiency of this factory's manufacturing lines and the mean efficiency of the null hypothesis population (established manufacturing lines in this industry).

Alternative hypothesis: There is a difference between the mean efficiency of this factory's manufacturing lines and the mean efficiency of the null hypothesis population.

 Step 1: Compute the probability of the mean efficiency of this sample given that the sample comes from the null hypothesis population of normally distributed efficiencies.

Formula for a *z* test:

$$z = \frac{\bar{X} - \mu}{\sigma_{\bar{X}}}.$$

Substitution of $\sigma_{\bar{X}}$:

$$\sigma_{\bar{X}} = \frac{\sigma}{\overline{)n}}.$$

Calculate the mean of the sample:

$$\bar{X} = \frac{(2.8 + 3.2 + 3.4 + 2.6)}{4} = 3.0 \text{ widgets per day per line}.$$

Applying the formula to our example:

$$z = \frac{3.00 - 2.80}{\dfrac{0.30}{\overline{)\,4}}} = \frac{0.20}{0.15} = 1.333333.$$

 Step 2: Evaluate the probability of obtaining this score due to chance.

Evaluate the *z*-obtained value based on alpha (α) = 0.05 and a two-tailed hypothesis.

From Table A, column c (in the Appendix): the probability of obtaining a *z*-obtained score of 1.333333 (look up 1.33 in the table) from the normally distributed null hypothesis population is 0.0918. However, the hypothesis we are evaluating in this case is two tailed, and the table only gives the area under the curve (or probability) for one tail, and thus we need to double that probability.

Double the probability for two-tailed evaluation:

$$0.0918 \times 2 = 0.1836.$$

Compare the two-tailed probability with our alpha level:

$$0.1836 > 0.05.$$

Our obtained probability of 0.1836 is greater than our α of 0.05, and thus we fail to reject the null hypothesis that suggests there is no difference in efficiency of widget production.

Alternatively, you could compare a z-critical value with your z-obtained value. When $\alpha = 0.05$ and your hypothesis is two tailed, you should use the z-critical value that lists 0.025 in column c (doubled, that would be 0.05). Thus, the z-critical values for this example would be ± 1.96.

To reject the null hypothesis, the absolute value of z obtained must be equal to or more extreme than (greater than) the z critical.

$$|z_{obtained}| \geq |z_{critical}|$$

$|1.33| < |1.96|$, so we fail-to-reject the null hypothesis.

How should we interpret these data in light of the efficiency of these new manufacturing lines relative to industry standards?

Based on a sample size of 4, these results suggest that about 18.36% of all samples of industry-standard manufacturing lines would have a similar mean efficiency. Thus, it is likely that this sample of line efficiencies comes from the normal null hypothesis population of manufacturing efficiencies, and it is not likely that this factory's manufacturing line efficiencies depart significantly from industry standards.

How would these results be reported in a scientific journal article?

If the journal required APA format, the results would be reported in a format something like this:

There was no significant difference between the mean efficiency of this factory's manufacturing lines and the mean efficiency of the null hypothesis population, $z(N = 4) = 1.33, p = .18$.

This formal sentence includes the dependent variable (mean efficiency), the independent variable (this factory vs. population of factories), as well as a statement about statistical significance, the symbol of the test (z), the sample size, the statistical value (1.33), and the approximate probability of obtaining this result simply due to chance (.18). The approximate probability is given in this case because APA format requires that you round to two digits.

When is it appropriate to use the z test?

1. When the experiment involves a single sample mean and the parameters of the corresponding null hypothesis population are known (μ and σ) *and*

2. When the sampling distribution is normally distributed, which is the case if

    a. The sample size is greater or equal to 30 ($n \geq 30$) or

    b. The null hypothesis population of raw scores is known to be normally distributed.

Unfortunately, it is not very often that scientists know the mean and the standard deviation of the population. It is a little more common to know the mean of the population but not the standard deviation. One example of this situation is when you can assume that the mean of the population is zero, but the standard deviation of the population is unknown. For example, you might assume that the average genetic relatedness of students in a random undergraduate statistics course is 0.00, but you could actually measure the relatedness through DNA sampling and determine whether the sample comes from the null hypothesis population of unrelated individuals. In this case, we "know" the mean of the null hypothesis population (0.00) but not the standard deviation, and we know the mean and standard deviation of the sample (of relatedness values determined by DNA sampling). We would like to compare the relatedness determined in our sample (the sample mean) to the population mean when we assume no relatedness (the population mean = 0.00), but we only know the standard deviation of the *sample,* not the *population.* In this case, a modification of the single-sample z test is appropriate, and it becomes the single-sample t test.

SINGLE-SAMPLE t TEST

In this situation, we want to compare a sample to a population, but we do not know the standard deviation of the population. The solution to this problem is to estimate the standard deviation of the population from the sample standard deviation. Remember that we used substitution to alter the single-*subject* z score to a single-*sample* z test. We will do that again to convert the single-sample z test to a single-sample t test, but this time we will substitute the sample standard deviation (s) for the population standard deviation (σ).

Comparison of Single-Sample z and Single-Sample t Formulas

$$z = \frac{\bar{X} - \mu}{\sigma_{\bar{X}}} = \frac{\bar{X} - \mu}{\dfrac{\sigma}{\sqrt{n}}}.$$

$$t = \frac{\bar{X} - \mu}{s_{\bar{X}}} = \frac{\bar{X} - \mu}{\dfrac{s}{\sqrt{n}}}.$$

So, just as $\sigma_{\bar{X}} = \dfrac{\sigma}{\sqrt{n}}$, it is true that $s_{\bar{X}} = \dfrac{s}{\sqrt{n}}$. It may be necessary in some problems to calculate the sample standard deviation (s) from raw scores. You will find the following formulas, from Chapter 4, helpful should you need to do that.

$$s = \sqrt{\frac{SS}{n-1}} \qquad SS = \sum X^2 - \frac{\left(\sum X\right)^2}{n}.$$

The t-Critical Values Are Dependent on Sample Size (n)

Because we are now going to be *estimating* σ from s, and the accuracy of that estimation is dependent on n (Chapter 4), our critical probability values are now going to vary with our sample size, unlike the z distribution in which the probabilities were independent of sample size. Specifically, we will use a new distribution called the t distribution. It will vary with **degrees of freedom** (*df*), which are related to sample size.

> **Degrees of freedom:** The number of scores that are "free to vary" when calculating a statistic. The remaining value or values are then fixed.

The degrees of freedom for the standard deviation are equal to n − 1. The reason for this is that there is an outside constraint. You learned in an earlier chapter that the sum of the deviations around the mean is always equal to zero. If that is the case and you know all of the scores except the last one, then the last score is fixed. It must be the score that is necessary to make the sum of the deviations around the mean equal to zero. Thus, there is a constraint on the last value. For example,

Score (X)	Sample Mean ($\bar{X}$)	Deviation
3	6	–3
5	6	–1
10	6	*FIXED VALUE = 4*
$\sum X = 18$		$\sum$ Deviations = 0

Note that the last deviation must be equal to 4 in order for the sum of the deviations to equal zero. Thus, the last score is not free to vary but must take the value that will lead to the sum of the deviations being equal to zero. This may seem a bit mysterious to you, but there is a reason for bringing up this issue. When we calculate t obtained, we must first calculate the standard deviation. Thus, we lose one degree of freedom each time we calculate s. Therefore, the single-sample t test has $n - 1$ degrees of freedom, and this is what determines your t-critical value. See the sidebar for more mathematical details on the degrees of freedom. In our example, two scores are free to vary, and hence there are two degrees of freedom. Let's move on to an example to clarify.

FYI: DEGREES OF FREEDOM

In statistics, the concept of "degrees of freedom" describes the number of values that are "free to vary" in an estimate of a parameter. In practice, we typically lose a degree of freedom for each parameter we estimate. For example, if we want to estimate the population standard deviation based on the sample deviation, then our degrees of freedom are equal to $n - 1$. In other words, all of the scores are free to vary except the last one, and that value is fixed. This explanation may be sufficient for many students, but for some, this concept may seem like magic without more mathematical logic.

The mathematical definition of "degrees of freedom" requires a reference to geometry. Cramer (1946) defines a number of degrees of freedom as the rank of a quadratic form. It can also be defined in terms of vectors, and that definition includes the number of components that are free to vary. Alternatively, it can be stated as the number of components that must be known before a vector can be fully determined. For a review of various mathematical definitions for degrees of freedom, see Good (1973).

Complete Single-Sample t *Test Example*

Let's repeat our example with AZT and birth weight in pigtailed macaques, but this time, let's assume that we do not know the standard deviation of the population. The population of birth weights is normally distributed, with a mean of 425 grams. A sample of four newborn macaques that were exposed to AZT had the following birth weights:

Subject	Birth Weight (g)
1	340
2	410
3	365
4	315

Null hypothesis: There is no difference in the birth weights of a sample of AZT-exposed pigtailed macaques and the null hypothesis population.

Alternative hypothesis: There is a difference in the birth weights of AZT-exposed pigtailed macaques and the null hypothesis population.

 Step 1: Compute the probability of the mean birth weight of this sample given that the sample comes from the null hypothesis population of normally distributed birth weights.

Formula for a *t* test:

$$t = \frac{\bar{X} - \mu}{s_{\bar{X}}}.$$

Substitution of $s_{\bar{X}}$:

$$s_{\bar{X}} = \frac{s}{\sqrt{n}}.$$

Calculate the mean of the sample:

$$\bar{X} = \frac{(340 + 410 + 365 + 315)}{4} = 357.5 \text{ grams}.$$

Calculate the sums of squares:

$$(\Sigma X)^2 = (340 + 410 + 365 + 315)^2 = 2{,}044{,}900.$$

$$\Sigma X^2 = (340)^2 + (410)^2 + (365)^2 + (315)^2 = 516{,}150.$$

Calculate the standard deviation of the sample:

$$s = \sqrt{\frac{SS}{n-1}} \qquad SS = \sum x^2 - \frac{(\sum x)^2}{n}.$$

$$SS = 516{,}150 - \frac{(1430)^2}{4} = 516{,}150 - 511{,}225 = 4{,}925.$$

$$s = \sqrt{\frac{4{,}925}{4-1}} = \sqrt{\frac{4{,}925}{3}} = \overline{) 1641.666667} = 40.51748594.$$

Applying the formula to our example:

$$t = \frac{357.5 - 425}{\dfrac{40.5}{\overline{)\,4}}} = \frac{-67.5}{20.25} = -3.33.$$

 Step 2: Evaluate the probability of obtaining this score due to chance.

Evaluate the *t*-obtained value based on alpha (α) = 0.05 and a two-tailed hypothesis. To evaluate your *t*-obtained value, you must use a new sampling distribution, which is the *t* distribution. You will find the *t* distribution table in the back of your book (Table B in the Appendix). To determine your *t*-critical value, you will need to know your alpha level (0.05), the number of tails you are evaluating (two in this case), and your degrees of freedom ($n - 1 = 4 - 1 = 3$). Because the critical value varies with degrees of freedom, a complete *t* distribution table would take up an entire textbook. Due to limited space, you will find only a limited number of *t*-critical values that are the most commonly used. You will not be able to determine the exact probability of obtaining each *t*-obtained value that you calculate. You will only know the exact probability of a given *t*-obtained value if it happens to be the same as a *t*-critical value that is listed in the Appendix. Thus, when performing hand calculations that you evaluate based on statistical tables, you are limited to determining only whether the probability of obtaining your statistical result is less than or greater than your alpha level. Fortunately, numerous software packages are available to calculate statistical tests, and these software packages contain the entire set of tables. Thus, you will be exposed to two of these software packages in future chapters so that you can interpret their probabilities simultaneously with examples worked out via manual calculations.

Using a t-*Critical Value to Evaluate Your Results*

Compare a *t*-critical value with your *t*-obtained value. When $\alpha = 0.05$, your degrees of freedom are equal to 3, and your hypothesis is two-tailed, you should use 3.182 as your *t*-critical value.

To reject the null hypothesis for a two-tailed test, the absolute value of *t* obtained must be equal to or more extreme than your *t* critical.

$$|t_{\text{obtained}}| \geq |t_{\text{critical}}|$$

$|-3.33| \geq |3.182|$, so we reject the null hypothesis.

How should we interpret these data in light of the risks of AZT on fetal development?

These results suggest that something less than 5% of all normal pregnancies have similar birth weights. Thus, it is unlikely that these birth weights come from the normal null hypothesis population of birth weights, and it is more likely that AZT affects birth weight. However, there is a chance ($< 5\%$) that these are normal birth weights that were not affected by AZT.

How would these results be reported in a scientific journal article?

If the journal required APA format, the results would be reported in a format something like this:

There was a significant difference between the mean birth weight of AZT-exposed pigtailed macaques ($M = 357.5$) and the null hypothesis population ($\mu = 425$), $t(3) = -3.33, p < .05$.

This formal sentence includes the dependent variable (birth weight), the independent variable (AZT exposed or non–AZT exposed), the direction of the effect as indicated by the means, as well as a statement about statistical significance, the symbol of the test (*t*), the degrees of freedom (rather than sample size), the statistical value (–3.33), and the probability compared to your alpha level (.05).

Second Complete Single-Sample t Test Example

A sociologist recorded the total number of arrests for a sample of rehabilitated inmates to determine whether they had been arrested fewer times than the general prison population. The population of arrests is normally distributed, with a mean of 7 arrests. The sample of six rehabilitated inmates had the following number of total arrests:

Rehabilitated Inmates	Number of Total Arrests
1	6
2	8
3	5
4	7
5	6
6	7

Null hypothesis: There is no difference in the total number of arrests for a sample of rehabilitated inmates and the null hypothesis population, or the rehabilitated inmates were arrested more often than the null hypothesis population.

Alternative hypothesis: Rehabilitated inmates were arrested fewer times than the null hypothesis population.

 Step 1: Compute the probability of the mean arrests of this sample given that the sample comes from the null hypothesis population of normally distributed arrests in the general prison population.

Formula for a *t* test:

$$t = \frac{\bar{X} - \mu}{s_{\bar{X}}} .$$

Substitution of $s_{\bar{X}}$:

$$s_{\bar{X}} = \frac{s}{\sqrt{n}} .$$

Calculate the mean of the sample:

$$\bar{X} = \frac{(6 + 8 + 5 + 7 + 6 + 7)}{6} = 6.5 \text{ arrests.}$$

Calculate the sums of squares:

$$(\Sigma X)^2 = (6 + 8 + 5 + 7 + 6 + 7)^2 = 1{,}521.$$

$$\Sigma X^2 = (6)^2 + (8)^2 + (5)^2 + (7)^2 + (6)^2 + (7)^2 = 36 + 64 + 25 + 49 + 36 + 49 = 259.$$

Calculate the standard deviation of the sample:

$$s = \sqrt{\frac{SS}{n - 1}} \qquad SS = \Sigma X^2 - \frac{(\Sigma X)^2}{n}.$$

$$SS = 259 - \frac{(39)^2}{6} = 259 - 253.5 = 5.5.$$

$$s = \sqrt{\frac{5.5}{6 - 1}} = \sqrt{\frac{5.5}{5}} = \sqrt{1.1} = 1.0488088.$$

Applying the formula to our example:

$$t = \frac{6.5 - 7}{\dfrac{1.0488088}{\sqrt{6.000000}}} = \frac{-0.5}{0.4281744} = -1.16677.$$

 Step 2: Evaluate the probability of obtaining this score due to chance.

Evaluate the *t*-obtained value based on alpha (α) = 0.05 and a one-tailed hypothesis. To evaluate your *t*-obtained value, you must use the *t* distribution (Table B in the Appendix). To determine your *t*-critical value, you will need to know your alpha level (0.05), the number of tails you are evaluating (one in this case), and your degrees of freedom ($n - 1 = 6 - 1 = 5$). You will not be able to determine the exact probability of obtaining each *t*-obtained value that you calculate. You will only know the exact probability of a given *t*-obtained value if it happens to be the same as a *t*-critical value that is listed in the Appendix.

Using a t-*Critical Value to Evaluate Your Results*

Compare a *t*-critical value with your *t*-obtained value. When α = 0.05, your degrees of freedom are equal to 5, and your hypothesis is one tailed, you should use 2.015 as your *t*-critical value.

To reject the null hypothesis for a one-tailed test, the t obtained must be equal to or more extreme than your t critical.

$$|t_{obtained}| \geq |t_{critical}|$$

$< |-2.015|$, so we fail-to-reject the null hypothesis.

How should we interpret these data in light of the effect of age on item recall for this particular memory test?

These results suggest that something more than 5% of all inmates are arrested this many times. Thus, it is possible that these rehabilitated inmates come from the normal null hypothesis population of inmates and that total arrest numbers do not affect the likelihood of rehabilitation.

How would these results be reported in a scientific journal article?

If the journal required APA format, the results would be reported in a format something like this:

There was no significant difference between the mean number of arrests of rehabilitated inmates and the null hypothesis population of all prison inmates, $t(5) = -1.17, p > .05$.

This formal sentence includes the dependent variable (number of arrests), the independent variable (rehabilitated inmates and all inmates), as well as a statement about statistical significance, the symbol of the test (t), the degrees of freedom (rather than sample size), the statistical value (-1.17), and the probability compared to your alpha level ($.05$).

When is it appropriate to use a single-sample t test?

1. When you have one sample, know the population mean but not the population standard deviation, and have an interest in comparing your sample mean to the population mean

2. When the sampling distribution is normally distributed, this is the case as earlier when
   a. The sample size is greater or equal to 30 ($n \geq 30$) or
   b. The null hypothesis population is known to be normally distributed.

How to Choose the Appropriate Test

Knowing how to choose the appropriate statistical test is arguably the most important skill you will learn in this course. Computers are better at calculating the tests than they are at determining the correct test for the design and measurement scale. Thus, you will need to hone your skills in choosing the appropriate test as you proceed through this course. To aid you in this development, we have provided a decision flowchart (Figure 8.2) that will become more complicated as you learn more statistical tests. Right now it is fairly simple, but becoming familiar with it early will be a great aid to you, and you can use it to solve the "choose the appropriate test" questions in the exercises for this chapter.

Figure 8.2

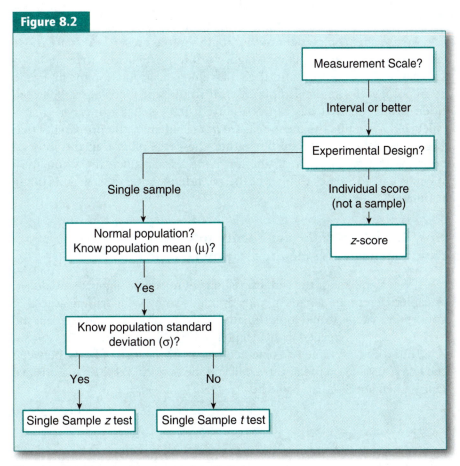

Flowchart for Choosing the Appropriate Test

CONFIDENCE INTERVALS: A VALUABLE USE OF THE SINGLE-SAMPLE *t* TEST FORMULA

In everyday life you use **confidence intervals**, but you don't think of them as confidence intervals. For example, if your spouse asks you how much money you spend at the race track, you might say, "$50, give or take $5." Essentially you are suggesting that your best guess is $50, but you are not entirely confident that $50 is the exact amount you spent, but you are willing to say that you spent $5 more or less of that amount. Mathematically, you are saying 50 ± 5, or 45 to 55 dollars. We use confidence intervals in the same way in statistics, but we assign probabilities to inform us of how confident we are that the range we have indicated includes the actual number of interest.

> **Confidence interval for the population mean:** Range of values with a calculated probability of containing the mean.

> **Confidence limits for the population mean:** The upper and lower values (or boundaries) surrounding the confidence interval.

We use a rearrangement of the single-sample *t* test to calculate those probabilities. For example, we might want to be 95% sure that the interval that we describe contains the true value (or population mean, in the terminology of a single-sample test). If so, we can calculate a 95% confidence interval that has a probability of 0.95 that the interval that is calculated contains the mean. We are *not* calculating the probability that the mean is in the interval but rather the probability that the interval contains the mean. This distinction is necessary since the population mean is a fixed, "true" thing, but the confidence interval is what is being estimated.

How do we derive the confidence interval from the single-sample *t* test formula?

If you think about the "critical regions of rejection" figures you looked at earlier, determining a mean, plus or minus some value, is similar to that concept. In fact, it would look something like Figure 8.3 if you overlaid the *t*-critical values associated with an alpha of 0.05 and two tails.

Thus, the true value of *t*-obtained has a 95% chance of being between the *t*-critical value on the negative tail and the *t*-critical value on the positive tail. Mathematically, that would be $-t_{0.025} \le t_{obtained} \le +t_{0.025}$.

However, $t_{obtained} = \dfrac{\bar{X} - \mu}{s_{\bar{X}}}$.

Figure 8.3

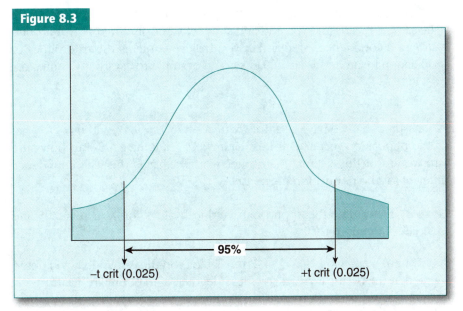

95%

−t crit (0.025)   +t crit (0.025)

Graphic Display of a 95% Confidence Interval

If you substitute the formula that we use for t obtained into the confidence limit formula above, you would have the following:

$$\frac{-t_{0.025} \leq \bar{X} - \mu \leq + t_{0.025}}{s_{\bar{X}}}.$$

If you solve for μ, then you get the following:

$$\bar{X} - s_{\bar{X}} \quad (t_{0.025}) \leq \mu \leq \bar{X} + s_{\bar{X}} (t_{0.025}).$$

This suggests that there is a 95% chance that the interval $\bar{X} + s_{\bar{X}} (t_{0.025})$ contains μ. Notice that the statement includes a plus and a minus after the sample mean.

$$\mu_{Lower} = \bar{X} - s_{\bar{X}} (t_{0.025}).$$

$$\mu_{Upper} = \bar{X} + s_{\bar{X}} (t_{0.025}).$$

This is the same as our earlier statement that you have your best guess (the sample mean) and you add and subtract some amount of error around it to get a confidence interval that you believe contains the true value (or population mean in this case). Let's do an example to clarify any confusion.

Confidence Interval Example

Assume that we have a sample with $\bar{X} = 75$, $s = 9.37$, and $n = 47$. What is the population mean? Our best estimate is, of course, 75. But how sure are we are of this estimate? In other words, what is the 95% confidence interval for the population mean, μ?

★ **Step 1:** Determine the appropriate t-critical values for your $\alpha = 0.05$ and degrees of freedom (df).

Your $df = n - 1 = 47 - 1 = 46$. From Table B (in the Appendix), you determine that the t-critical values for these degrees of freedom are ±2.021.

★ **Step 2:** Calculate the standard error.

$$s_{\bar{X}} = \frac{s}{\sqrt{n}}.$$

$$s_{\bar{X}} = \frac{9.37}{\sqrt{47}} = 1.3667.$$

★ **Step 3:** Insert your values into the formula.

$$\mu_{\text{Lower}} = 75 - 1.3667(2.021) = 72.24.$$

$$\mu_{\text{Upper}} = 75 + 1.3667(2.021) = 77.76.$$

★ **Step 4:** Interpret your confidence interval.

There is a 95% probability that the interval from 72.24 to 77.76 contains the population mean. So, our best estimate of the population mean is 75, and we are very sure (95% sure) that the interval between 72.2 and 77.8 contains the population mean. This would be quite different than saying that the mean was 75 and that we were quite sure (95% sure) that the interval between 55 and 95 contained the population mean, a much larger confidence interval indicating much less confidence that we really know what interval contains the population mean. This larger confidence interval would be the result of a smaller sample size, a larger sample standard deviation, or both.

Usefulness of Confidence Intervals

We have worked through an example involving a 95% confidence interval, but it is possible to calculate confidence intervals of 90% or 99% or any confidence interval that you like. What changes is the *t*-critical value that you insert into the equation. To determine the confidence interval with 99% confidence means that your confidence interval will actually be larger than the size of a 95% confidence interval. This seems counterintuitive at first, but if you think carefully, you will realize that to be 99% sure that the interval contains the mean, you have make the interval larger. Thus, if we have calculated a 99% confidence interval in the example above, the *t*-critical value would have been 2.704 rather than the 2.021 that we used. Thus, we end up adding a larger number and subtracting a larger number, which increases the size of our interval. A number of statisticians advocate using confidence intervals rather than hypothesis tests to determine the likelihood that a sample mean came from the null hypothesis population. While we recognize the logic behind this approach, our strategy will be to teach you both the confidence interval approach as well as the hypothesis tests since hypothesis testing is still prevalent in the scientific literature.

EFFECT SIZES AND POWER

Effect size is a standardized measure of the difference between two (or more) group means; it is the difference in means divided by the shared standard deviation of two or more groups. For a population with a known standard deviation (σ), you can use Cohen's *d*. If testing a sample with no known population parameters, you can estimate power by using Hedges's *g* (covered in Chapter 9):

$$\textbf{Cohen's } \boldsymbol{d} \quad \frac{\mu_1 - \mu_2}{\sigma}.$$

Once you have calculated your effect size, you can use a free program to calculate the power of your test. One that we have often recommended to students is "G*Power," which can be downloaded at www.psycho .uni-duesseldorf.de/aap/projects/gpower/.

The program can be used a priori to predict adequate sample size, based on desired effect size and desired power, or post hoc to calculate power based on calculated effect size and standard error.

A priori, after typing in the desired effect size (based on previous research or current hypotheses) and desired power (usually 0.8 or your research won't be funded), total sample size, along with t_{crit} and actual power, appears in the middle and bottom of the screen.

Post hoc power is determined by inputting effect size and sample size and hitting "Calculate." This process is especially useful after you have completed a preliminary study and are proposing a full study. Given the effect size of your pilot study, you can predict how many subjects are necessary to obtain significant results with a specific likelihood (for example, 80%).

Other Resources for Calculating Power

If you have an experimental design that does not fit with the options for calculating power with G*Power, you can use the following website to find more free resources on the Internet for calculating statistics, including power: http://statpages.org/.

SUMMARY

In this chapter, we have introduced our first hypothesis tests, those that test hypotheses having to do with single-sample designs when compared to underlying populations. We further developed the single-sample z statistic to compare a sample mean to a population mean by introducing hypothesis testing to that concept. We went on to make this concept more useful by relaxing the (admittedly uncommonly satisfied) assumption of a known population standard deviation by allowing the estimation of the population standard deviation from the sample standard deviation: the single-sample t test. This required a different sampling distribution for calculating the resulting probabilities of the null hypothesis, and we introduced the t distribution, which is influenced by sample size or actually degrees of freedom. Finally, we applied the single-sample t test to a very useful descriptive statistic called the confidence interval. In the next chapter, we will introduce a new design, two samples, and describe how the t test can be adapted to test hypotheses with this design.

CHAPTER 8 HOMEWORK

Define the following terms.

1. What is the null hypothesis population?

2. Why is it called the null hypothesis population?

Answer these short-answer questions.

3. If we must estimate the population standard deviation to determine the significance of the sample results, the appropriate inference test is:

4. List three assumptions of the single-sample z test.

Multiple-Choice Questions

5. In order to reject the null hypothesis, which of the following decision rules must be true?

   A. $t_{obt} < t_{crit}$

   B. $t_{obt} > t_{crit}$

   C. $t_{obt} < t_{crit}$

   D. $t_{obt} > t_{crit}$

6. The sampling distribution of t varies with _____.

   A. μ

   B. χ_-

   C. σ

   D. s

   E. df

7. As the df increases, the t distribution _____.

   A. gets larger

   B. gets smaller

   C. stays the same

   D. approaches the normal distribution

8. With which statistical test(s) do we determine the probability of getting the obtained results or those more extreme based on the likelihood of them occurring due to chance alone?

   A. z test

   B. t test

C. None of the above

D. A and B

9. If we do not know the population standard deviation but we do know the population mean and we want to compare it to a sample, the appropriate statistical test is _____.

   A. *t* test

   B. binomial distribution test

   C. *z* test

   D. all of the above

10. How does the single-sample *t* test differ from the single-sample *z* test?

    A. With the *t* test we estimate μ

    B. With the *z* test we estimate μ

    C. The *z* distribution is a family of curves

    D. The *t* distribution is a family of curves

    E. None of the above

11. How are the degrees of freedom for the single-sample *t* test determined?

    A. *n*

    B. *n* – 1

    C. *n* – 2

    D. *df* – 1

12. A social scientist believes that the average number of pages in a professional journal in the social sciences has declined in the past few decades because of high publication costs. A previous study indicated that between 1920 and 1970, the average page length of 70 social science journals was 115. To investigate this belief, she randomly samples 41 social science journals and records their page lengths. The journals have a mean of 111 pages and a standard deviation of 12.4 pages. Using $\alpha = .01$ (one-tailed), what do you conclude?

13. Assess the following statistics (accept or reject the null hypothesis) under the conditions described. Use the most powerful and appropriate test statistic.

    A. You have a control population of 10,000; $\mu = 67.6$, $\sigma = 5.5$; one-tailed, $\alpha = 0.01$; experimental sample: $N = 30$, $\chi\_ = 72.1$, $s = 15.6$.

    B. You have a normally distributed population of 35; $\mu = 1.36$, $\sigma = 1.17$; two-tailed, $\alpha = 0.05$; sample $n = 3$, $\chi\_ = 2.04$, $s = 0.15$.

Choose the most appropriate test to analyze the following story problems. You do not need to calculate the statistic but merely list the appropriate test for analyzing this scenario.

14. A researcher is interested to see if there is a relationship between aging and depression. On a standardized depression test, the general population averages age 40. The researcher tests a random sample of 45 individuals who are all older than age 70 on the depression test and obtains a mean of 44.5. Does there appear to be a relationship between depression and aging?

15. A health club is interested to find out if individuals younger than age 28 spend more or less time in the gym than the rest of the general population. The average time spent in the gym per week for the average population is 3.5 hours with a standard deviation of 2.7. The mean hours of time spent in the gym of the 35 individuals younger than age 28 who participated in the study was 5.6 with a standard deviation of 1.8. Test for statistical significance.

Answer Questions 16 to 20 based on the following story problem.

A random sample of 80 college students was selected to participate in an experimental anxiety reduction program designed to decrease test anxiety. After completing the anxiety reduction program, participants completed a battery of tests designed to measure anxiety level. The mean on the anxiety measures for the sample was 317 points, with a standard deviation of 37. According to the researcher's vast data on anxiety, the mean anxiety level for college students who have not completed the anxiety reduction program is 310 points.

16. Compute the appropriate test statistic.

17. What is the critical value of the test statistic?

18. Given these data, what would you conclude?

19. Calculate a 95% and a 99% confidence interval.

20. Which confidence interval is broader and why is that the case?

Answer Questions 21 to 25 based on the following story problem.

Mrs. Bell has been teaching kindergarten in King County for 20 years. One day while talking with her daughter, who is a senior at the University of Washington, she mentioned that she believed that the average 5-year-old girl is taller today than when she first started teaching. To investigate this belief, her daughter (an aspiring statistician) randomly sampled 41 five-year-old girls at Northgate mall and recorded their heights. The mean height was 46 inches and a standard deviation of 3.5 inches. A local census completed in 1980 (30 years ago) shows that the mean height of 5-year-old girls was 43.75 inches.

21. Is the teacher's hypothesis directional or nondirectional?

22. Compute the appropriate test statistic.

23. What is the critical value of the test statistic?

24. Given these data, what would you conclude?

25. If Mrs. Bell's daughter had the standard deviation of heights for 5-year-olds from the 1980 census, what would have been the appropriate test statistic to compute?

CHAPTER 9

Two-Sample Tests

If you have any interest in knowing how to statistically demonstrate that there is a significant difference between your control group and your experimental group, or in the before and after effects of your educational program, then this chapter should help. In fact, the statistical tests you are about to learn are (arguably) the most common tests reported in professional journals!

In the last chapter, you learned how to evaluate hypotheses for tests when you had one sample and known population parameters. While those tests are powerful, population parameters can be difficult to obtain. Here we introduce the two-sample tests, where you will compare two samples that came from the same population, rather than comparing a single sample to a population. The samples may be completely independent from one another (between-groups design) or related in some way (within-groups design).

Independent or between-groups designs are those in which subjects are randomly selected from a population and are randomly assigned to either the control or experimental conditions. Subjects only serve in one condition.

Dependent or within-groups designs are those in which subjects are randomly selected from a population and serve in more than one condition (such as "before" vs. "after" some treatment) or subjects are matched into pairs and one subject in each pair serves in each condition. Because either the same subjects or subjects that are similar to one another in some significant way serve in the within-groups design experiments, the amount of variation due to nuisance factors is minimized in these designs. When the variability that is not of interest to the researchers is minimized, the power of the experiment is increased. Remember from Chapter 7 that power is the (highly desirable) property that measures the likelihood that a given experimental design will be able to detect a real effect if a real effect exists. Thus, within-groups designs are more powerful than between-groups designs, and we will introduce the test to analyze those designs first. We provide examples for each statistical test, so you learn how to calculate the tests, as well as how to interpret the results, what they look like in computer software output, and how you present the results for professional publications. Next, we return to the concepts of power and effect size, which are topics that are critical for interpretation of your results and are often required for publication.

Notice that with each chapter, we are now logically building the flowchart for choosing the appropriate test. The order of the chapters is meant to provide a logical extension to your statistical knowledge and to allow you to make sense of the myriad tests used to analyze data.

PAIRED *t* TEST (CORRELATED GROUPS *t* TEST)

The paired *t* test (also called the "correlated groups" *t* test) is used when you have two samples and a within-groups design. This design is also called a dependent or repeated-groups design. Both the name of the statistical test and the name of the research design can vary a great deal from book to book and between different statistical software packages. You can navigate this confusion by having a conceptual understanding of what the test is doing. This statistical test requires that you have met one of the following *experimental design* conditions:

1. You have two measures on the same subjects ("before" and "after" measures are common). See the example in Table 9.1.

or

2. You have two separate samples but the subjects in each are *individually* matched so that there are similar subjects in each group (but not the same subjects in each group). For example, you might match

subjects on age and sex, so that you have a 36-year-old woman in your
control group and a 36-year-old woman in your experimental group,
a 28-year-old man in your control group and a 28-year-old man in your
experimental group, and so on. This can also be done by placing one
identical twin in the control group and the other twin in the experimental group or by any matching of *individuals* that is an attempt
(see Table 9.2). Note that the matching must be pairwise, so that you
can literally compare the scores of the twins side by side. You'll see
why this is important when you see the formula for the paired *t* test.

Table 9.1		
Subject	Score Before Treatment	Score After Treatment
1	50	55
2	52	58
3	44	48
4	42	41
5	49	56

Example of "Before" and "After" Pairing Using the Same Subjects in Each "Paired"
Sample

Table 9.2		
Twin Pair	Twin 1 = Control Group	Twin 2 = Experimental Group
A	10	8
B	12	10
C	21	19
D	18	15

Example of a "Paired" Design in Which the Actual Subjects in Each Sample Are
Different but Are "Matched" for Characteristics That They Have in Common (Genetics
in This Example)

Paired *t* Test Calculation

The calculation of the paired *t* test statistic comes from a modification of the single-sample *t* test. However, now we first calculate a difference score for each pair of scores in our two samples and treat those
difference scores as a single sample that will be compared to the mean

difference score (μ_D) of the null hypothesis population. The mean difference score of the null hypothesis population is assumed to be zero ($\mu_D = 0$)—that is, no difference between our samples or no effect of our independent variable (see Table 9.3). The mean difference score for the paired samples is a measure of any effect of our treatment. If our treatment does not have an effect, then there will not be a difference between the two groups, and the mean difference score will be zero (or close to it), like the mean difference score of the null hypothesis population. However, if the treatment does have an effect, it will increase or decrease the scores from the control condition and therefore produce a mean difference score greater or less than zero.

Thus, we can calculate the sum of the difference scores (ΣD), the sum of the squared difference scores (ΣD^2), and the mean difference score ($\bar{D}$) and the standard deviation of the difference scores (s_D). These difference scores become our single sample of raw scores that we contrast with a null hypothesis population mean of no difference between subjects ($\mu_D = 0$), and we estimate the standard deviation of the population difference scores (σ_D) based on the sample difference scores, just as we did in the single-sample t test. Because we are using an estimate of the population difference scores based on the sample but assume that we know the population mean ($= 0$), we use the t distribution to evaluate our t obtained value.

For comparison, we first present the formula for the single-sample t test with which we are familiar. Then we present the modification of this formula where we have replaced the mean of the sample with the mean of the difference scores (our new sample) and the standard error of the sample means with the standard error of the mean differences. The idea is that the formula for the paired t test is really just the formula for the single-sample t

Table 9.3			
Twin Pair	**Twin 1 = Control**	**Twin 2 = Experimental**	**Difference Score**
A	10	8	2
B	12	10	2
C	21	19	2
D	18	15	3
			$\Sigma D = 9$
			$\bar{D} = 9/4 = 2.25$

Example of Difference Score Calculations for a Paired t Test

test if you consider the difference scores to be your "single sample." That's the secret of the paired *t* test. Also, remember that *n* in the paired *t* test formula refers to the number of difference scores or the number of pairs of data points, not the total number of data points.

Formula for a Single-Sample *t* Test (Review)

$$t_{\text{obtained}} = \frac{\bar{X} - \mu}{s_{\bar{X}}} = \frac{\bar{X} - \mu}{\frac{s}{\sqrt{n}}}.$$

Paired *t* Test Formula

$$t_{\text{obtained}} = \frac{\bar{D} - \mu_D}{s_{\bar{D}}} = \frac{\bar{D} - \mu_D}{\frac{s_D}{\sqrt{n}}} = \frac{\bar{D} - \mu_D}{\sqrt{\frac{SS_D}{n(n-1)}}}.$$

Remember that under the null hypothesis, $\mu_D = 0$, so our formula becomes

$$t_{\text{obtained}} = \frac{\bar{D}}{s_{\bar{D}}} = \frac{\bar{D}}{\frac{s_D}{\sqrt{n}}} = \frac{\bar{D}}{\sqrt{\frac{SS_D}{n(n-1)}}}.$$

A Quick Example: A One-Tailed Paired t Test

A researcher interested in employee satisfaction and productivity measured the number of units produced by employees at a plant before and after a company-wide pay raise occurred. The researcher hypothesized that production would be higher after the raise compared to before the raise. Assume that the difference scores are normally distributed and let $\alpha = 0.05$.

Null hypothesis: There is no difference in the number of units produced before and after the raise, or the number of units was higher before the raise.

Alternative hypothesis: The number of units produced was higher after the raise.

 Step 1: Compute the probability of the mean differences of this sample given that the sample comes from the null hypothesis population of difference scores where $\mu_D = 0$.

Calculate the difference scores and the intermediate numbers for the SS formula:

Participants	Before	After	Difference Score	D²
1	7	7	*0*	*0*
2	4	5	*−1*	*1*
3	8	9	*−1*	*1*
4	8	9	*−1*	*1*
5	6	6	*0*	*0*
6	6	6	*0*	*0*
7	5	5	*0*	*0*
8	5	4	*+1*	*1*
9	7	7	*0*	*0*
			$\sum D = -2$	$\sum D^2 = 4$
			$n = 9$	
			$\bar{D} = 2/9 = -0.2222222$	

Calculate the standard deviation of the sample:

$$SS_D = \sum D^2 - \frac{\left(\sum D\right)^2}{n} = 4 - \frac{(-2)^2}{9} = 4 - 0.4444444 = 3.5555555\cdot$$

$$s_D = \sqrt{\frac{SS_D}{(n-1)}} = \sqrt{\frac{3.5555}{9-1}} = \sqrt{0.4444375} = 0.6666614\,.$$

Apply the formula to our example:

$$t = \frac{\bar{D} - \mu_D}{\dfrac{s_D}{\sqrt{n}}}\,.$$

$$t = \frac{-0.2222222 - 0}{\frac{0.6666614}{\sqrt{9}}} = \frac{-0.2222}{0.22222046} = -0.9999079\,.$$

 Step 2: Evaluate the probability of obtaining this score due to chance.

Evaluate the *t*-obtained value based on alpha (α) = 0.05 and a one-tailed hypothesis. To evaluate your *t*-obtained value, you must use the *t* distribution (Table B in the Appendix) as you did in the last chapter. To determine your *t*-critical value, you need to know your alpha level (0.05), the number of tails you are evaluating (one in this case), and your degrees of freedom ($n - 1 = 9 - 1 = 8$). Compare the *t*-critical value with your *t*-obtained value. When $\alpha = 0.05$, your degrees of freedom are equal to 8, and your hypothesis is one-tailed, you should use 1.86 as your *t*-critical value.

To reject the null hypothesis for a *t* test, the *t*-obtained must be equal to, *or more extreme than,* the *t*-critical value. Be sure to also check that the effect is in the correct direction (correct based on the hypothesis).

$$|t_{\text{obtained}}| \geq |t_{\text{critical}}|$$

$|-0.99| < |-1.86|$, so we *fail to reject* the null hypothesis.

How should we interpret these data in light of the effect of the raise on productivity?

These results suggest that more than 5% of the time, you would obtain this number of units regardless of whether it was after a raise. Thus, it is likely that the difference in these production values (before and after the raise) comes from the normal null hypothesis population of difference scores. However, remember that there is a chance that there is a real effect of raises on productivity that we have not detected in this analysis.

Complete Example

A sociologist is interested in the decay of long-term memory compared to the number of errors in memory that an individual made after 1 week and after 1 year for a specific crime event. Participants viewed a videotape of a bank robbery and were asked a number of specific questions about the video 1 week after viewing it. They were asked the same questions 1 year after seeing the video. The number of memory errors was recorded for each participant at each time period. The researchers asked whether or not there was a significant difference in the number of errors in the two time periods. Assume that the difference scores are normally distributed and let $\alpha = 0.05$.

Null hypothesis: There is no difference in the number of errors made at 1 week and at 1 year.

Alternative hypothesis: There is a difference in the number of errors made at 1 week and at 1 year.

 Step 1: Compute the probability of the mean differences of this sample given that the sample comes from the null hypothesis population of difference scores where $\mu_D = 0$.

Calculate the difference scores and the intermediate numbers for the SS formula:

Subject	One Week	One Year	Difference Score	D²
1	5	7	*−2*	*4*
2	4	5	*−1*	*1*
3	6	9	*−3*	*9*
4	8	9	*−1*	*1*
5	6	6	*0*	*0*
6	5	6	*−1*	*1*
7	4	5	*−1*	*1*
8	5	4	*+1*	*1*
9	7	7	*0*	*0*
			$\Sigma D = -8$ $n = 9$ $\bar{D} = -8/9 = -0.8888$	$\Sigma D^2 = 18$

Calculate the standard deviation of the sample:

$$SS_D = \sum D^2 - \frac{\left(\sum D\right)^2}{n} = 18 - \frac{(-8)^2}{9} = 18 - 7.1111 = 10.8888 \cdot$$

$$S_D = \sqrt{\frac{SS_D}{(n-1)}} = \sqrt{\frac{10.8888}{9-1}} = \sqrt{1.36110} = 1.16666 \cdot$$

Apply the formula to our example:

$$t = \frac{\bar{D} - \mu_D}{\dfrac{S_D}{\sqrt{n}}}.$$

$$t = \frac{-0.8888 - 0}{\dfrac{1.16666}{\sqrt{9}}} = \frac{-0.8888}{0.388887} = -2.2855 \cdot$$

 Step 2: Evaluate the probability of obtaining this score due to chance.

Evaluate the t-obtained value based on alpha (α) = 0.05 and a two-tailed hypothesis. To evaluate your t-obtained value, you must use the t distribution (Table B) as you did in Chapter 8. To determine your t-critical value, you need to know your alpha level (0.05), the number of tails you are evaluating (two in this case), and your degrees of freedom ($n - 1 = 9 - 1 = 8$). Compare the t-critical value with your t-obtained value. When $\alpha = 0.05$, your degrees of freedom are equal to 8, and your hypothesis is two-tailed, you should use 2.306 as your t-critical value.

To reject the null hypothesis for a two-tailed test, the absolute value of t obtained must be equal to, or more extreme than, the t-critical value.

$$|t_{obtained}| \geq |t_{critical}|$$

$|-2.2855| \leq |2.306|$, so we *fail to reject* the null hypothesis.

How should we interpret these data in light of the effect of time on the number of memory errors?

These results suggest that more than 5% of the time, you would obtain this number of memory errors regardless of whether it was after 1 week or 1 year. Thus, it is likely that these memory error differences come from the normal null hypothesis population of difference scores. However, remember that there is a chance that there is a real effect of time on memory errors that we have not detected in this analysis.

Results if you use Microsoft Excel to calculate the *t* test:

	One Week	One Year
Mean	5.555556	6.444444
Variance	1.777778	3.027778
Observations	9	9
Pearson Correlation	0.742315	
Hypothesized Mean Difference	0	
Df	**8**	
t Stat	**−2.28571**	
P(T<=t) one-tail	0.025804	
t Critical one-tail	1.859548	
P(T<=t) two-tail	0.051609	
t Critical two-tail	**2.306006**	

t-Test: Paired Two Sample for Means

This is the output that you get from Excel when you type in these data for 1 week and 1 year. Note that Excel calls this test "*t*-Test: Paired Two Sample for Means." This wording is slightly different from what we have been using, but it is describing the same analysis. We have bolded the numbers that are comparable to the numbers we just manually calculated or looked up in the table. Note that the *t*-obtained value (Excel calls it "t Stat") is identical to ours, except they have carried more digits (−2.28571). The "t Critical two-tail" is also the same (2.306006), but again they have carried more digits. The degrees of freedom are the same (8). But they have also provided you with the estimated probability of obtaining a *t* value (t Stat) of −2.28571, which is 0.051609. Since 0.051609 > 0.05, we clearly cannot reject our null hypothesis. Thus, we fail to reject the null hypothesis. These results suggest that 5.1609% of the time you would obtain this number of memory errors regardless of whether it was after 1 week or 1 year. Knowing the estimated probability for each and every *t*-obtained value (not just the *t*-critical values) is one of the major advantages of using a computer to calculate your analyses.

Knowing the calculated probability in this case ($p = .052$) raises another issue: .052 is pretty close to our critical probability of .050. Our results suggest that there is a 5.16% chance that any differences were due to chance, as opposed to our critical level of 5.00%. Many researchers would present these results as "significant" even though the probability technically exceeds the critical level. The point is that there is nothing magical about the critical alpha value of 0.05, 0.01, or whatever value is chosen: Nothing special happens between $p = .052$ and $p = .049$. We have to use some common sense and knowledge of the system being studied to decide whether we are convinced that a real effect was demonstrated. This is an important point in critical thinking about statistics.

Results if you use SPSS to calculate the *t* test:

		Paired Differences			95% Confidence Interval of the Difference		t	df	Sig. (2-tailed)
		Mean	Std. Deviation	Std. Error Mean					
					Lower	Upper			
Pair 1	ONEWEEK— ONEYEAR	**−.8889**	**1.1667**	**.3889**	−1.7857	7.890E-03	**−2.286**	**8**	.052

Paired Samples Test

We have placed the numbers that are comparable to our manual calculations in bold. Once again, you see that calculated probability (SPSS calls Sig. 2-tailed) is greater than our alpha level of 0.05, and thus we must assume that these results could occur by chance and not necessarily as a result of time since the event.

How would these results be reported in a scientific journal article?

If the journal required American Psychological Association (APA) format, the results would be reported in a format something like this:

There was no significant difference in the number of errors made at 1 week and at 1 year, $t(8) = -2.29, p > .05$.

This formal sentence includes the dependent variable (number of errors), the independent variable (1 week vs. 1 year), as well as a statement about statistical significance, the symbol of the test (t), the degrees of freedom (8), the statistical value (−2.29), and the estimated probability of obtaining this result simply due to chance (> .05).

Another Complete Example

An animal behaviorist is concerned about the effects of nearby construction on the nesting behavior (trips to nest per hour) of endangered dusky seaside sparrows in Florida. She knows that the quality of the nesting territory's habitat will also influence this nesting behavior, so she picks seven pairs of nests, each with the same territory quality (say, density of seed plants), one nest of the pair near the construction and one in an undisturbed location, for a total of 14 nest observations. Assume that the difference scores are normally distributed and let $\alpha = 0.05$.

Null hypothesis: There is no difference in the rate of nest visits made at "construction" nests and "undisturbed" nests.

Alternative hypothesis: There is a difference in the rate of nest visits between the two locations. No specific direction of the difference is suggested.

 Step 1: Compute the probability of the mean differences of this sample given that the sample comes from the null hypothesis population of difference scores where $\mu_D = 0$.

Calculate the difference scores and the intermediate numbers for the SS formula:

Matched Pair	Undisturbed	Construction	Difference Score	D^2
1	5.4	3.2	*2.2*	*4.84*
2	4.1	3.5	*0.6*	*0.36*
3	9.7	7.1	*2.6*	*6.76*
4	8.4	6.8	*1.6*	*2.56*
5	6.0	6.4	*−0.4*	*0.16*
6	6.0	4.5	*1.5*	*2.25*
7	7.9	7.6	*0.3*	*0.09*
			$\sum D = 8.4$	$\sum D^2 = 17.02$
			$n = 7$	
			$\bar{D} = 8.4/7 = 1.2$	

Rate of Nest Visits Per Hour

Calculate the standard deviation of the sample:

$$SS_D = \sum D^2 - \frac{(\sum D)^2}{n} = 17.02 - \frac{(8.4)^2}{7} = 17.02 - 10.08 = 6.94.$$

$$S_D = \sqrt{\frac{SS_D}{(n-1)}} = \sqrt{\frac{6.94}{7-1}} = \sqrt{1.15667} = 1.07548.$$

Apply the formula to our example:

$$t = \frac{\bar{D} - \mu_D}{\frac{S_D}{\sqrt{n}}}.$$

$$t = \frac{1.2 - 0}{\left(\frac{1.07548}{\sqrt{7}}\right)} = \frac{1.2}{0.40649} = 2.95210.$$

 Step 2: Evaluate the probability of obtaining this score due to chance.

Evaluate the *t*-obtained value based on alpha (α) = 0.05 and a two-tailed hypothesis. To evaluate your *t*-obtained value, you must use the *t* distribution (Table B) as you did in Chapter 8. To determine your *t*-critical value, you need to know your alpha level (0.05), the number of tails you are evaluating (two in this case), and your degrees of freedom ($n - 1 = 7 - 1 = 6$). Compare the t-critical value with your t-obtained value. When $\alpha = 0.05$, your degrees of freedom are equal to 6, and your hypothesis is two-tailed, you should use ±2.447 as your t-critical value.

To reject the null hypothesis for a two-tailed test, the absolute value of t obtained must be equal to, or more extreme than, the t-critical value.

$$|t_{\text{obtained}}| \geq |t_{\text{critical}}|$$

$|2.952| \geq |2.447|$, so we *reject* the null hypothesis.

How should we interpret these data in light of the effect of construction on the rate of nest visits?

These results suggest that *less than 5%* of the time, you would obtain this rate of nest visits regardless of whether it was near or not near to the construction site. Thus, it is likely that the rate of nest visits near the construction site *does not* come from the same underlying population of scores as the nest site visits away from the construction site, and therefore, the difference scores in this example do not represent a null hypothesis population of difference scores. However, remember that there is a chance (however small) that there is, in reality, no real effect of the construction on nest site visits, and our conclusion is in error.

Results if you use Microsoft Excel to calculate the *t* test:

	Undisturbed	Construction
Mean	6.785714	5.585714
Variance	3.784762	3.284762
Observations	7	7
Pearson Correlation	0.838487	
Hypothesized Mean Difference	0	
Df	**6**	
t Stat	**2.952067**	
P(T<=t) one-tail	0.012772	
t Critical one-tail	1.943181	
P(T<=t) two-tail	0.025544	
t Critical two-tail	**2.446914**	

t-Test: Paired Two Sample for Means

This is the output that you get from Excel when you type in these data for undisturbed and construction sites. Once again, we have bolded the numbers that are comparable to the numbers we just manually calculated or looked up in the table. Note again that the *t*-obtained value is identical to ours, except they have carried more digits. The "t Critical two-tail" is also the same, but again they have carried more digits. The degrees of freedom are the same (6). But they have also provided the calculated probability of obtaining a *t* value (t Stat) of 2.952067, which is 0.025544. Since 0.025544 < 0.05, we can reject our null hypothesis. Thus, we reject the null hypothesis. These results suggest that 2.5544% of the time, you would be making a mistake in concluding that there was a "real effect" of the construction on nest site visits. Traditionally, we are willing to accept a 5% probability of making such a mistake. Again, knowing the calculated probability for each *t*-obtained value (not just the *t*-critical values) is one of the major advantages of using a computer to calculate your analyses.

Results if you use SPSS to calculate the *t* test:

		Paired Differences					t	df	Sig. (2-tailed)
		Mean	Std. Deviation	Std. Error Mean	95% Confidence Interval of the Difference				
					Lower	Upper			
Pair 1	UNDISTUR— CONSTRUC	1.2000	1.0755	.4065	.2053	2.1947	2.952	6	.026

Paired Samples Test

Once again, we have placed the numbers that are comparable to our manual calculations in bold. And once again, you see that calculated probability (.026: SPSS calls it "Sig. 2-tailed") is less than our alpha level of 0.05, and thus we must assume that these results are unlikely due to chance and thus are more likely a result of time since the event.

How would these results be reported in a scientific journal article?

If the journal required APA format, the results would be reported in a format something like this:

There was a significant difference in the rate of nest visits between the undisturbed location ($M = 6.79$, $SD = 1.94$) and the construction location ($M = 5.59$, $SD = 1.81$), $t(6) = 2.95$, $p = .03$.

This formal sentence includes the dependent variable (rate of nest visits), the independent variable (two different locations), the direction of the effect as evidenced by the reported means, as well as a statement about statistical significance, the symbol of the test (t), the degrees of freedom (6), the statistical value (2.95), and the estimated probability of obtaining this result simply due to chance (.03).

When is it appropriate to use a paired *t* test?

1. When you have two samples and a within-groups design

2. When the sampling distribution is normally distributed, which, as you should recall, is satisfied when

    a. The sample size is greater or equal to 30 ($n \geq 30$) or

    b. The null hypothesis population is known to be normally distributed.

3. When the dependent variable is on an interval or ratio scale

INDEPENDENT *t* TEST

In the previous section, we described a situation where your two conditions contain either the same subjects or subjects that have been individually matched on an important characteristic that might potentially influence the outcome of your results and is not interesting to you (age or body weight are potential examples of characteristics that you might match subjects on). This is referred to as a within-groups design. Now we turn to the situation where you actually have two completely different (independent) groups of subjects that you want to compare to determine if they are significantly different from one another: a between-groups design. The classic example of this is when you have a sample and you randomly assign half of your subjects to the control condition and the other half to the experimental treatment condition. In this situation, we wish to compare the means of the two conditions/groups. We can no longer assume that we know a population mean (as we did when we assumed that the $\mu_D = 0$ in the paired *t* test), and we must develop a new sampling distribution.

Sampling Distribution of the Difference Between the Means

To test for the potential statistical significance of a true difference between sample means, we need a sampling distribution of the difference between sample means $(\bar{X}_1 - \bar{X}_2)$. This would be a sampling distribution that will provide us with the probability that the difference between our two sample means $(\bar{X}_1, \bar{X}_2)$ differs from the null hypothesis population of sample mean differences: a population in which there is no difference between samples or, restated, the independent variable has no effect. The sampling distribution of the difference between the means can be created by taking all possible sample sizes of n_1 and n_2, calculating the sample means, and then taking the difference of those means. If you do this repeatedly for all of the possible combinations of your sample sizes, then you end up with a family of distributions of differences between the two means when they are randomly drawn from the same null hypothesis population. Choice of the specific distribution to be used in a problem depends on the degrees of freedom, as always.

The sampling distribution of the difference between sample means has the following characteristics:

1. If the null hypothesis population of scores is normally distributed, then the population of differences between the sample means will also be normally distributed.

2. The mean of the sampling distribution of the difference between sample means $\left(\mu_{\bar{X}_1 - \bar{X}_2}\right)$ will be equal to $\mu_1 - \mu_2$ (just as $\mu_{\bar{X}} = \mu$).

3. The standard deviation of the sampling distribution of the difference between sample means will be equal to the square root of the sum of each sample variance, or $\sqrt{\sigma_{\bar{X}_1}^2 + \sigma_{\bar{X}_2}^2}$.

Estimating Variance From Two Samples

Consider for a moment that the formulas we've used in the past contained one measure of variability. In the case of the single-sample z test, we used the population standard deviation, while with the single-sample t test, we estimated the population standard deviation with the sample standard deviation. When we performed the paired t test earlier in this chapter, we calculated the standard deviation of the sample of difference scores, so we were able to come up with one measure of sample variability. Now we have two independent samples, and each has a sample standard deviation. This forces us to determine which standard deviation best represents the true variability. In fact, what we do is to estimate the true population variability (or variance, σ^2) by taking the average variance (s^2) of our samples but weighted by their respective sample sizes. Remember, as we learned in earlier chapters, sample size or degrees of freedom affects the accuracy of our variance estimates, so an estimate from a sample with a large sample size would be more accurate than an estimated variance from a smaller sample. So we need to weight our average variance by the respective sample sizes of each sample. In using this approach, we are going to make a new assumption—that the sample variances are estimating the same underlying population variance, the variance of the null hypothesis population. Later in this chapter, we will have to make sure that our two sample variances are the same, within the bounds of random sampling error. This is referred to as the homogeneity of variance assumption.

Formula for Weighted Variance:

$$s_w^2 = \frac{df_1 s_1^2 + df_2 s_2^2}{df_1 + df_2}.$$

Substituting the equation for degrees of freedom and for variance:

$$= \frac{(n_1 - 1)(SS_1/n_1 - 1) + (n_2 - 1)(SS_2/n_2 - 1)}{(n_1 - 1) + (n_2 - 1)}.$$

Rearranging to simplify:

$$= \frac{SS_1 + SS_2}{n_1 + n_2 - 2}.$$

We have shown the formula in three different ways. The first way is the most intuitive way to present the average variance of the two samples when it is weighted by the sample size or, more specifically, by the appropriate degrees of freedom for each sample. The degrees of freedom are used because we are estimating the population variation from a sample, and thus one degree of freedom is lost each time we do that (one for each sample). The second formula actually plugs in the appropriate formulas for variance and degrees of freedom into the first formula, and the last formula is created by algebraic rearrangement into a simplified version. The first formula may be the best one to use if you are obtaining each sample variance from your calculator directly from these raw data or if you are given either variance or standard deviation of each sample in a problem. The last formula would be best if you have already calculated the sums of squares (SS) for each group.

Derivation of the Independent *t* Test Formula From the Single-Sample *t* Test Formula

Formula for a single-sample *t* test (for review):

$$t_{\text{obtained}} = \frac{\bar{X} - \mu}{\frac{s}{\sqrt{n}}}.$$

Conceptual formula for an independent *t* test:

$$t_{\text{obtained}} = \frac{(\bar{X}_1 - \bar{X}_2) - \mu_{\bar{X}_1 - \bar{X}_2}}{\sqrt{\frac{\sigma_1^2}{n_1} + \frac{\sigma_2^2}{n_2}}} = \frac{(\bar{X}_1 - \bar{X}_2) - \mu_{\bar{X}_1 - \bar{X}_2}}{\sqrt{\sigma^2(1/n_1 + 1/n_2)}}.$$

Calculation (practical) formula to use for an independent *t* test:

$$t_{\text{obtained}} = \frac{\bar{X}_1 - \bar{X}_2}{\sqrt{s_w^2(1/n_1 + 1/n_2)}}.$$

The formula for an independent *t* test is derived by assuming the mean of the sampling distribution or differences between means is zero for the null hypothesis population and by using the average variance divided by each sample size. The square root in the denominator of the independent *t* test formula is there not only to take the square root of the sample size (as you did when you calculated a single-sample *t* test) but also because you are working in squared units (variance), and we must take the square root of the variance to get back to the standard deviation.

Variations of the Independent *t* Test Formula

All-purpose formulas:

$$t_{\text{obtained}} = \frac{\bar{X}_1 - \bar{X}_2}{\sqrt{s_w^2(1/n_1 + 1/n_2)}} = \frac{\bar{X}_1 - \bar{X}_2}{\sqrt{\left(\frac{SS_1 + SS_2}{n_1 + n_2 - 2}\right)(1/n_1 + 1/n_2)}}.$$

To create the second variation of the formula, we simply substituted the formula for s_w^2 directly into the independent *t* test formula.

Formula to be used only when $n_1 = n_2$:

$$t_{\text{obtained}} = \frac{\bar{X}_1 - \bar{X}_2}{\sqrt{\left(\frac{SS_1 + SS_2}{n(n-1)}\right)}}.$$

A Quick One-Tailed Example of an Independent t *Test*

A researcher breeds rats for nine generations but only breeds the rats that perform very well in a maze (few errors) to each other (called "maze-bright" rats) and also breeds rats that perform very poorly in a maze (many errors) to one another ("maze-dull" rats). After nine generations, is there a significant decrease in the number of errors of the maze-bright rats?

Null hypothesis: There is no difference in the number of errors made by the maze-bright and the maze-dull rats or there are more errors in the maze-bright rats.

Alternative hypothesis: There is a significant decrease in the number of errors of the maze-bright rats.

 Step 1: Compute the probability that each of the sample means comes from the null hypothesis population of differences between means.

Calculate the means and the intermediate numbers for the SS formula:

Maze-Bright Rats	Maze-Dull Rats
2	6
3	4
4	5
3	3
4	6
2	
$\sum X_{\text{BRIGHT}} = 18$	$\sum X_{\text{DULL}} = 24$
$\left(\sum X_{\text{BRIGHT}}\right)^2 = 18^2 = 324$	$\left(\sum X_{\text{DULL}}\right)^2 = 24^2 = 576$
$\sum X_{\text{BRIGHT}}^2 = 58$	$\sum X_{\text{DULL}}^2 = 122$
$\bar{X}_{\text{BRIGHT}} = \dfrac{18}{6} = 3$	$\bar{X}_{\text{DULL}} = \dfrac{24}{5} = 4.8$

Number of Errors in a Maze by Ninth-Generation Rats

Calculate the SS of each sample:

$$SS_{\text{BRIGHT}} = \sum X^2 - \frac{\left(\sum X\right)^2}{n} = 58 - \frac{(18)^2}{6} = 58 - \frac{324}{6} = 4.$$

$$SS_{\text{DULL}} = \sum X^2 - \frac{\left(\sum X\right)^2}{n} = 122 - \frac{(24)^2}{5} = 122 - \frac{576}{5} = 6.8.$$

Apply the SS formula of the independent t test to our example:

$$t_{\text{obtained}} = \frac{\bar{X}_1 - \bar{X}_2}{\sqrt{\left(\frac{SS_1 + SS_2}{n_1 + n_2 - 2}\right)(1/n_1 + 1/n_2)}}.$$

$$t = \frac{3 - 4.8}{\sqrt{\left(\frac{4 + 6.8}{6 + 5 - 2}\right)(1/6 + 1/5)}} = \frac{-1.8}{\sqrt{(1.2)(.3666)}} = -2.714$$

★ **Step 2:** Evaluate the probability of obtaining this score due to chance.

Evaluate the t-obtained value based on alpha (α) = 0.05 and a one-tailed hypothesis. To evaluate your t-obtained value, you must use the t distribution (Table B). To determine your t-critical value, you need to know your alpha level (0.05), the number of tails you are evaluating (one in this case), and your degrees of freedom (df). The degrees of freedom for an independent t test are $(n_1 - 1) + (n_2 - 1)$ or $n_1 + n_2 - 2$. Thus, the df for this problem are $6 + 5 - 2 = 9$. Compare the t-critical value with your t-obtained value. When $\alpha = 0.05$, your degrees of freedom are equal to 9, and your hypothesis is one-tailed, you should use 1.833 as your t-critical value.

$-2.7414 > -1.833$, so we *reject* the null hypothesis.

How should we interpret these data in light of the effect of time on the number of memory errors?

These results suggest that less than 5% of the time, you would obtain this difference in the number of errors if the breeding had no effect. Thus, it is not very likely that these error differences come from the normal null hypothesis population. However, there is a chance that you could get a difference this large between two means that is purely due to chance, but that chance is less than 5%. Note that you can refer back to the means to determine that the effect was in the correct direction. Specifically, maze-bright rats made 3 errors on average and maze-dull rats made 4.8 errors on average. Not only is the *t*-obtained value more extreme than the *t*-critical value, but the direction of the effect is as the researcher predicted.

Complete Example

A researcher breeds rats for nine generations but only breeds the rats that perform very well in a maze (few errors) to each other (maze-bright rats) and also breeds rats that perform very poorly in a maze (many errors) to one another (maze-dull rats). After nine generations, is there a significant difference in the number of errors in the two groups of rats?

> *Null hypothesis:* There is no difference in the number of errors made by the maze-bright and the maze-dull rats or the differences are due to chance.

> *Alternative hypothesis:* There is a difference in the number of errors made by each group.

Evaluate the *t*-obtained value based on alpha (α) = 0.05 and a two-tailed hypothesis. To evaluate your *t*-obtained value, you must use the *t* distribution (Table B). To determine your *t*-critical value, you need to know your alpha level (0.05), the number of tails you are evaluating (two in this case), and your degrees of freedom (*df*). The degrees of freedom for an independent *t* test are $(n_1 - 1) + (n_2 - 1)$ or $n_1 + n_2 - 2$. Thus, the *df* for this problem are $6 + 5 - 2 = 9$. Compare the *t*-critical value with your *t*-obtained value. When $\alpha = 0.05$, your degrees of freedom are equal to 9, and your hypothesis is two-tailed, you should use 2.262 as your *t*-critical value.

$$|-2.714| \geq |2.262|, \text{ so we } reject \text{ the null hypothesis.}$$

How should we interpret these data in light of the effect of time on the number of memory errors?

These results suggest that *less than 5%* of the time, you would obtain this difference in the number of errors if the breeding had no effect. Thus, it is not very likely that these error differences come from the normal null hypothesis population. However, there is a chance that you could get a difference this large between two means that is purely due to chance, but that chance is less than 5%.

Results if you use Microsoft Excel to calculate the *t* test:

	Bright	**Dull**
Mean	**3**	**4.8**
Variance	0.8	1.7
Observations	**6**	**5**
Pooled Variance	1.2	
Hypothesized Mean Difference	0	
Df	**9**	
t Stat	**−2.713602101**	
P(T<=t) one-tail	0.011928192	
t Critical one-tail	1.833113856	
P(T<=t) two-tail	0.023856384	
t Critical two-tail	**2.262158887**	

t-Test: Two-Sample Assuming Equal Variances

This is the output you get from Excel when you type in these data for maze-bright and maze-dull rats. Note that Excel calls this test "*t*-Test: Two-Sample Assuming Equal Variances." This wording is slightly different from what we have been using, but it is describing the same analysis, and that will be even clearer after we have discussed the assumptions of the independent *t* test. We have bolded the numbers that are comparable to the numbers we just manually calculated or looked up in the table. Note that the *t*-obtained value (Excel calls it "t Stat") is identical to ours, except they have carried more digits (−2.713602101). The "t Critical two-tail" is also the same (2.262158887), but again they have carried more digits. The degrees of freedom are the same (9). In addition, they have provided the calculated probability of obtaining a *t* value (t Stat) of −2.713602101, which is 0.023856384. Since 0.023856384 < 0.05, we clearly *reject* our null hypothesis. These results suggest that 2.3856384% of the time, you would obtain this difference by chance if breeding had no effect. Again, knowing the calculated probability for each and every *t*-obtained value (not just the *t*-critical values) is one of the major advantages of using a computer to calculate your analyses.

Results if you use SPSS to calculate the *t* test:

		Levene's Test for Equality of Variances		t-test for Equality of Means					95% Confidence Interval of the Difference	
		F	Sig.	t	df	Sig. (2-tailed)	Mean Difference	Std. Error Difference	Lower	Upper
ERRORS	Equal variances assumed	1.255	.292	**−2.714**	**9**	.024	**−1.8000**	.6633	−3.3005	−.2995
	Equal variances not assumed			−2.616	6.903	.035	−1.8000	.6880	−3.4315	−.1685

Independent Samples Test

We have placed the numbers that are comparable to our manual calculations in bold. Once again, you see that calculated probability (SPSS calls it Sig. 2-tailed) is less than our alpha level of 0.05, and thus we must assume that these results would be *unlikely* simply due to chance.

How would these results be reported in a scientific journal article?

If the journal required APA format, the results would be reported in a format something like this:

There was a significant difference in the number of errors made by maze-dull ($M = 4.8$, $SD = 1.30$) and maze-bright rats ($M = 3.0$, $SD = 0.89$), $t(9) = −2.71$, $p = .02$.

This formal sentence includes the dependent variable (number of errors), the independent variable (two groups of rats), as well as a statement about statistical significance, the direction of the effect as evidence by the means, the symbol of the test (t), the degrees of freedom (9), the statistical value (−2.71), and the estimated probability of obtaining this result simply due to chance (.02).

Another Complete Example

Psychology experiments frequently need to be very careful about the effects of past experience on the outcome of their studies. One common population for psychology research is college students enrolled in psychology courses. But it is traditional to use only freshmen or sophomores who are enrolled in introductory courses so that the researcher avoids the possibility of a student anticipating the study's goals after having had more advanced psychology courses. But does classroom experience in psychology really affect the outcome of the experiment? A researcher develops a simple assessment for response to visual stimuli, and the number of correct (matching) results is scored for a group of five sophomore introductory psychology course students and six psychology seniors. Is there a significant difference in the scores of the two groups of students?

Null hypothesis: There is no difference in the scores of introductory and advanced psychology students or the differences are due to chance.

Alternative hypothesis: There is a difference in the scores of each group.

 Step 1: Compute the probability that each of the sample means comes from the null hypothesis population of differences between means.

Calculate the means and the intermediate numbers for the SS formula:

Introductory	Advanced
9	8
8	9
7	8
9	6
6	10
9	
$\sum X_{Intro} = 39$	$\sum X_{Adv} = 50$
$(\sum X_{Intro})^2 = 39^2 = 1521$	$(\sum X_{Adv})^2 = 50^2 = 2500$
$\sum X_{Intro}^2 = 311$	$\sum X_{Adv}^2 = 426$
$\bar{X}_{Intro} = \dfrac{39}{5} = 7.80$	$\bar{X}_{Adv} = \dfrac{50}{6} = 8.33$

Scores for Introductory and Advanced Students

Calculate the SS of each sample:

$$SS_{\text{Intro}} = \sum X^2 - \frac{\left(\sum X\right)^2}{n} = 311 - \frac{(39)^2}{5} = 311 - \frac{1521}{5} = 6.80.$$

$$SS_{\text{Adv}} = \sum X^2 - \frac{\left(\sum X\right)^2}{n} = 426 - \frac{(50)^2}{6} = 426 - \frac{2500}{6} = 9.33.$$

Apply the SS formula of the independent t test to our example:

$$t_{\text{obtained}} = \frac{\bar{X}_1 - \bar{X}_2}{\sqrt{\left(\frac{SS_1 + SS_2}{n_1 + n_2 - 2}\right)(1/n_1 + 1/n_2)}}.$$

$$t = \frac{7.80 - 8.33}{\sqrt{\left(\frac{6.80 + 9.33}{5 + 6 - 2}\right)(1/5 + 1/6)}} = \frac{-0.53}{\sqrt{(1.79222)(.36667)}} = -0.6538.$$

 Step 2: Evaluate the probability of obtaining this score due to chance.

Evaluate the t-obtained value based on alpha (α) = 0.05 and a two-tailed hypothesis. To evaluate your t-obtained value, you must use the t distribution (Table B). To determine your t-critical value, you need to know your alpha level (0.05), the number of tails you are evaluating (two in this case), and your degrees of freedom (*df*). The degrees of freedom for an independent t test are $(n_1 - 1) + (n_2 - 1)$ or $n_1 + n_2 - 2$. Thus, the *df* for this problem are $5 + 6 - 2 = 9$. Compare the t-critical value with your t-obtained value. When $\alpha = 0.05$, your degrees of freedom are equal to 9, and your hypothesis is two-tailed, you should use 2.262 as your t-critical value.

$|-0.6538| < |2.262|$, so we *fail to reject* the null hypothesis.

How should we interpret these data in light of the effect of experience on scores?

These results suggest that *greater than 5%* of the time, you would obtain this difference in the scores if experience level had no effect. Thus, it is likely that these differences come from the normal null hypothesis population. In terms of our experiment, freshmen and seniors react the same way: Classroom experience does not have an effect on this response to visual stimuli. However, there is always a chance that there is a significant real effect that you did not detect.

Results if you use Microsoft Excel to calculate the *t* test:

	Introductory	Advanced
Mean	**7.800000**	8.333333
Variance	1.700000	1.866667
Observations	**5**	**6**
Pooled Variance	1.792593	
Hypothesized Mean Difference	0	
Df	**9**	
t Stat	**−0.657843**	
P(T<=t) one-tail	0.263552	
t Critical one-tail	1.833114	
P(T<=t) two-tail	0.527105	
t Critical two-tail	**2.262159**	

t-Test: Two-Sample Assuming Equal Variances

This is the output you get from Excel when you type in these data for introductory and advanced psychology students. We have again bolded the numbers that are comparable to the numbers we just manually calculated or looked up in the table. Note again that the *t*-obtained value is identical to ours, except they have carried more digits. The "t Critical two-tail" is also similar to the table except for differences due to rounding (2.262158887), and the same as in the first example since the degrees of freedom are the same (9). In addition, they have provided the calculated probability of obtaining a *t* value (t Stat) of −0.657843, which is 0.527105. Since 0.527105 > 0.05, we clearly *fail to reject* our null hypothesis. These results suggest that 52.7105% of the time, you would obtain this difference in errors if psychology experience had no effect.

Results if you use SPSS to calculate the *t* test:

		Levene's Test for Equality of Variances		t-test for Equality of Means						
		F	Sig.	t	Df	Sig. (2-tailed)	Mean Difference	Std. Error Difference	95% Confidence Interval of the Difference	
									Lower	Upper
DATA	Equal variances assumed	.008	.929	**−.658**	**9**	.527	−.5333	.8107	−2.3673	1.3007
	Equal variances not assumed			−.661	8.785	.526	−.5333	.8069	−2.3655	1.2989

Independent Samples Test

We have placed the numbers that are comparable to our manual calculations in bold. Once again, you see that that calculated probability (SPSS calls it Sig. 2-tailed) is greater than (*much* greater than) our alpha level of 0.05, and thus we must assume that these results would be *likely* simply due to chance.

How would these results be reported in a scientific journal article?

If the journal required APA format, the results would be reported in a format something like this:

> There is no significant difference between the scores of introductory and advanced psychology students, $t(9) = -.66, p = .53$.

This formal sentence includes the dependent variable (scores), the independent variable (introductory vs. advanced psychology students), as well as a statement about statistical significance, the symbol of the test (*t*), the degrees of freedom (9), the statistical value (−.66), and the probability of obtaining this result simply due to chance (.53).

When is it appropriate to use an independent *t* test?

1. When you have two samples and a between-groups design

2. When the sampling distribution is normally distributed. Again, this is satisfied when

   a. The sample size is greater or equal to 30 ($N \geq 30$) or

   b. The null hypothesis population is known to be normally distributed.

3. When the dependent variable is on an interval or ratio scale

4. When the variances of the two groups are the same or are homogeneous. The homogeneity of variance assumption (HOV) requires that the variances of the underlying populations are equal or, in practical terms, not significantly different from one another. We can test this assumption by examining the sample variances. For now, we will use a simple rule of thumb to decide whether the variances are similar enough to be considered homogeneous: If the $\frac{\text{larger variance}}{\text{smaller variance}} \leq 4$, then the HOV assumption is met (okay) and you can proceed with the independent *t* test. Note that there are more formal ways to test this assumption besides this rule of thumb. Interested students should explore information on Levene's test for more information. However, if the assumption is not met, you cannot proceed with the independent *t* test without some corrections or alterations. One possible solution is to run an unequal variances test rather than an independent *t* test. For example, Welch's *t* test is an adaptation of Student's *t* test intended for use with two samples having possibly unequal variances.

POWER AND TWO-SAMPLE TESTS: PAIRED VERSUS INDEPENDENT DESIGNS

Because high sample variability decreases power, and more variability would be expected from between-groups designs where different individuals are in each condition or group, the independent *t* test is less powerful than the paired *t* test. The paired *t* test gains power by controlling nuisance factors such as the age and temperament of a subject since the subjects are either the same in both conditions or match on characteristics that might influence the outcome of the study but are not of interest to the researcher. Thus, the researcher is more likely to control variability in the within-groups design.

There are several major drawbacks to the within-groups (or paired) design. One occurs when subjects drop out of the experiment prior to experiencing all of the conditions. This creates a situation where the researcher has one data point for the subject but lacks the other data point and thus cannot calculate a difference score for the subject. There are two options in this situation. The first option is to drop those subjects with missing data from your analysis, but this reduces your power significantly if you have a large number of missing data points relative to your sample size. The second option is to treat the two samples as independent and calculate an independent t test, and thus equal ns are not required and difference scores are not the basis for the test.

A second major drawback of the within-groups design is what generally is called experience effects: any research design in which novelty or experience with the assessment tool would bias the results. For example, any study in which the participants must be naive to the assessment to be able to provide a meaningful response would preclude the use of the within-groups design. You would require that both your control or pretest group and your treatment group be different individuals who had never before experienced the assessment itself.

EFFECT SIZES AND POWER

Effect size is a standardized measure of the difference between two (or more) group means; it is the difference in means divided by the shared standard deviation of two or more groups. If testing a sample with no known population parameters, you can estimate the effect size by using Hedges's g:

$$\text{Hedges's } g = \frac{\bar{X}_1 - \bar{X}_2}{S_p}.$$

Where S_p (when sample sizes are equal) $= \sqrt{\dfrac{s_x^2 + s_y^2}{2}}.$

Where S_p (when sample sizes are unequal) $= \sqrt{\dfrac{(n_x - 1)s_x^2 + (n_y - 1)s_y^2}{(n_x - 1) + (n_y - 1)}}.$

Once you have calculated your effect size, you can use a free program to calculate the power of your test. One that we have often recommended to students is "G*Power," which can be downloaded at www.psycho .uni-duesseldorf.de/aap/projects/gpower/.

The program can be used a priori to predict adequate sample size, based on desired effect size and desired power, or post hoc to calculate power based on calculated effect size and standard error.

A priori, after typing in the desired effect size (based on previous research or current hypotheses) and desired power (usually 0.8 or your research won't be funded), total sample size, along with t_{crit} and actual power, appears in the middle and bottom of the screen.

Post hoc power is determined by inputting effect size and sample size and hitting "Calculate." This process is especially useful after you have completed a preliminary study and are proposing a full study. Given the effect size of your pilot study, you can predict how many subjects are necessary to obtain significant results with a specific likelihood (e.g., 80%).

Other Resources for Calculating Power

If you have an experimental design that does not fit with the options for calculating power with G*Power, you can use the following website to find more free resources on the Internet for calculating statistics, including power: http://statpages.org/.

Notice that the flowchart (see page 175) does not yet tell you what to do if you do not meet the assumptions of your statistical tests. In fact, there are some options when you have two samples, but you do not meet the assumptions for parametric tests. The paired t test can be replaced with a nonparametric test called the Wilcoxon signed ranks test, while the independent t test can be replaced with a nonparametric test called the Mann-Whitney U test. We will discuss these nonparametric options in greater detail in Chapter 14. Another option is to perform a log transformation on your data to determine if the mathematical transformation will result in a normal distribution.

SUMMARY

In this chapter, you have covered statistical tests for designs in which you have two samples and lack information about the underlying population. Because we rarely know population parameters such as the mean and standard deviation (μ and σ), the two-sample tests tend to be used more often than the one-sample tests. Two-sample tests are less powerful because we are forced to estimate characteristics of the population, whereas one-sample tests rely on known measures of the population. In the next chapter, we will extend the concept of an independent t test or two-sample between-groups design to a situation where we have three samples and a between-groups design.

Excel Step-by-Step: Step-by-Step Instructions for Using Microsoft Excel 2003 or 2007 to Run t Tests

1. Your first step will be to open Microsoft Excel and type the raw data into a spreadsheet (data listed on page 176). It is helpful to type the column headers so that your output will be labeled later. Note that the participants who received caffeine in the first treatment are different

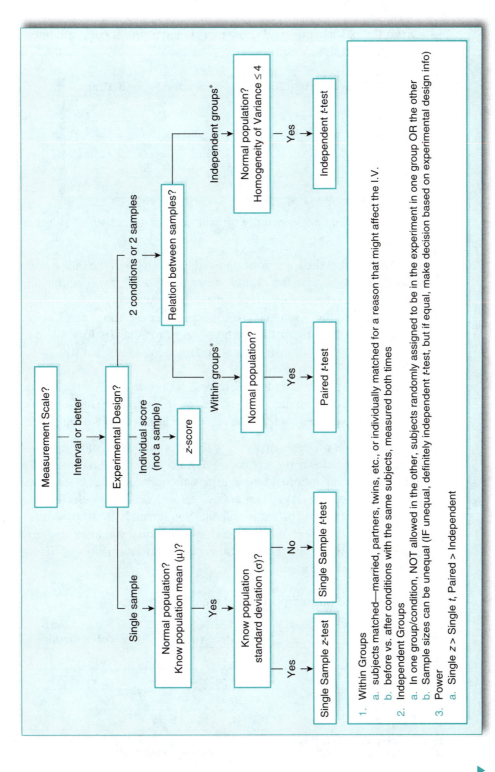

Overview of Single-Sample and Two-Sample Tests: Expanded Flowchart

Measurement Scale?

Interval or better

Experimental Design?

Single sample

Normal population?
Know population mean (μ)?

Yes

Know population
standard deviation (σ)?

Yes → Single Sample z-test

No → Single Sample t-test

**Individual score
(not a sample)**

z-score

2 conditions or 2 samples

Relation between samples?

Within groups*

Normal population?

Yes → Paired t-test

Independent groups*

Normal population?
Homogeneity of Variance ≤ 4

Yes → Independent t-test

1. Within Groups
   a. subjects matched—married, partners, twins, etc., or individually matched for a reason that might affect the I.V.
   b. before vs. after conditions with the same subjects, measured both times
2. Independent Groups
   a. In one group/condition, NOT allowed in the other, subjects randomly assigned to be in the experiment in one group OR the other
   b. Sample sizes can be unequal (IF unequal, definitely independent t-test, but if equal, make decision based on experimental design info)
3. Power
   a. Single z > Single t, Paired > Independent

175

from the participants who received caffeine in the second treatment (between-groups design).

Caffeine First Times	Caffeine Second Times
30	18
32	15
30	17
33	20
32	18

2. Once your data are entered, you will need to calculate an independent *t* test.

For Excel 2003: To find the *t* test you will need, you can go to the built-in data analysis function. You'll find the option under the "Tools" menu, and at the bottom of the list that pops up, you should see "Data Analysis."

If you do not see the "Data Analysis" option under the Tools menu, select "Add Ins" under the Tool menu. Check the box next to "Analysis TookPak" and click OK. Follow any further instructions that the computer gives you.

For Excel 2007: To get to the data analysis option, click on the "DATA" tab and the "Data Analysis" tool will be in the "Analysis" section to the far right of the screen. Once you find the Data Analysis tool, the rest of the instructions are the same.

3. After you click on "Data Analysis," a list of possible statistical tests will pop up. Page down the list until you find the appropriate test. We want to do an independent *t* test, which Excel calls "*t*-Test: Two Sample Assuming Equal Variances." After you have selected the appropriate test, the program will take you through the steps to complete the test. Note that Excel also has an unequal variances test. You can use this test if you meet the assumptions of normality but do not meet the variance assumption.

Here is example output from an independent *t* test:

	Control Group	Alcohol Group
Mean	2.7	4
Variance	1.566666667	1.555555556
Observations	10	10
Pooled Variance	1.561111111	
Hypothesized Mean Difference	0	

	Control Group	Alcohol Group
df	18	
t Stat	−2.326544946	
P(T<=t) one-tail	0.015932925	
t Critical one-tail	1.734063062	
P(T<=t) two-tail	0.031865849	
t Critical two-tail	2.100923666	

t-Test: Two-Sample Assuming Equal Variances

SPSS Step-by-Step: Step-by-Step Instructions for Using SPSS to Run an Independent *t* Test

1. Your first step will be to open SPSS and select the option that allows you to type in new data.

2. This will open a page called "Variable View." To confirm that, look at the tab at the bottom left of the page. There should be two tabs, and one will say "Variable View" (the one you are in now) and the other will say "Data View."

3. Now you need to establish your variables for SPSS. Make a variable that codes for your independent groups by typing in the word *group* in the first box of row 1. Now name your dependent variable and call it *dv* and type that into box 1 in row 2. By default, SPSS will consider each of these variables to be numeric, and for these purposes, all of the default codes will work perfectly. However, keep in mind that this is where you can change some of your options to allow for alphabetical data, define coding in your variables, and so on. For example, you could define your group coding to be 1 for the control group and 2 for the experimental treatment group.

4. Click on the Data View tab now. You should see that the variable names you entered in Variable View have now appeared at the top of this spreadsheet. Now you can enter your raw data:

Group	Dv
2	18
2	15

(Continued)

(Continued)

Group	Dv
2	17
2	20
2	18
1	30
1	32
1	30
1	33
1	32

5. From the SPSS menu, you should now select "Analyze," then "Compare means," and finally "Independent t-test." This will open up a new pop-up window with your variables listed on the left-hand side. Select your dependent variable and use the arrow in the middle of the pop-up to move it to the right-hand side of the pop-up box as your "test variable." Now move your "group" variable the same way over to the "group variable" box.

6. Now you will need to define the groups for your grouping variable. Click this option and type 1 for Group 1 and 2 for Group 2.

7. Once your groups are defined and you are back to the independent t test pop-up box, you can just hit "OK" to run the test.

8. You should get output for an independent t test assuming equal variances and for an independent t test where you do not need to assume equal variances. SPSS will also include Levene's test, which tells you whether or not your variances are homogeneous. If they are homogeneous, you should report the independent t test assuming equal variances. If your variables are not homogeneous and are heterogeneous, then you report the unequal variances t test. In addition to the t value and significance level (p value), the output will also include descriptive statistics such as the difference between your means, standard error of the means, your degrees of freedom, and the confidence interval.

CHAPTER 9 HOMEWORK

Provide a short answer for the following questions.

1. Why are paired (or correlated) designs more powerful than independent designs?

2. What are the assumptions for the paired *t* test?

3. What are the assumptions for the independent *t* test?

4. List three ways that you can meet the assumption of normality.

5. Why does the independent *t* test require the assumption of homogeneity of variance but the paired *t* test does not require this assumption?

6. How should you handle a situation where you have paired design and two conditions but there are some missing data for one of your two conditions?

7. What do you do if you do not meet the assumptions of the paired *t* test or the independent *t* test?

8. A researcher records the number of positive early childhood memories that can be recalled by five individuals who grew up in military families to the number of memories of individuals who grew up in nonmilitary families. The number of memories is normally distributed in each group. Using $\alpha = 0.05$ (two-tailed), what do you conclude?

Military family	18	25	17	20	23
Nonmilitary family	20	23	26	30	28

9. A sociologist is interested in whether or not race affects the likelihood that the average person will "shoot" a potential criminal in a computer simulation. Participants are required to make quick decisions about whether to "shoot" or not, and they are shown a variety of images of people. Some of the images are of people with a weapon and some of them are people holding nonviolent objects. Eight participants are randomly sampled for the study. The psychologist records the number of errors (shooting someone holding a nonviolent object) the participants made based on race (African American or Caucasian). The number of errors is normally distributed. The following data are recorded.

Participant	1	2	3	4	5	6	7	8
African American	28	29	25	30	25	27	28	24
Caucasian	25	28	22	30	26	24	25	22

Using $\alpha = 0.05$ (two-tailed), what do you conclude?

10. Professor Jones is intensely curious about differences in testing situations and wondered if students tended to make better scores on her tests depending on whether the test was taken on a Monday morning or a Friday morning. Her exams have always been normally distributed. From a group of 19 similarly talented students, she randomly selected some to take a test on Friday and others to take it on Monday. The scores by groups were as follows:

Monday	Friday
89.8	87.3
90.2	87.6
98.1	87.3
91.2	91.8
88.9	86.4
90.3	86.4
99.2	93.1
94.0	89.2
88.7	90.1
83.9	

Using $\alpha = 0.05$ (two-tailed), what do you conclude?

For Questions 11 to 12, you should choose the most appropriate and powerful test. Support your answers and list assumptions you are making. Do not try to perform the calculations.

11. Extensive research has been done on the subject of birth order. Data on this research show that first-born children develop different characteristics than later-born children. For example, first-born children tend to be more responsible and self-disciplined than later-born children. A researcher is interested in finding out if first-born children tend to be more confident and have higher self-esteem than later-born siblings. A random sample of 31 first-born children and 35 later-born children were given a self-esteem test. The standard deviation for the first-born children is 1.34, and the standard deviation for the later-born children is 2.15. Test whether birth order affects self-esteem.

12. A researcher tested a new medicine to see if it would be effective in lowering blood pressure. Two samples of participants from a normally distributed population were matched for medical history and initial blood pressure readings. Fifteen randomly selected participants were run through the

experimental condition in which they received the new drug. The other 15 randomly selected individuals participated in the control group and received a placebo. Participants receiving the drug showed a lower blood pressure. Test whether the drug had a statistically significant effect on blood pressure. Variances are homogeneous.

Answer Questions 13 to 19 using the following story problem.

A marketer is interested in how an antismoking campaign affects the smoking habits of teenagers. The researcher samples 50 students from a local area high school and asks them how many cigarettes they smoked. After the antismoking campaign has run for a year, the researcher polls the same 50 students and records the exact number of cigarettes smoked after the campaign.

13. State the null hypothesis for this study.

14. State the alternative hypothesis for this study.

15. Is this a directional or nondirectional hypothesis?

16. Is this research an independent groups (between subjects) or a repeated measures/paired design (within subjects)? Why?

17. What are the independent and dependent variables?

18. Which type of measurement scale do the data from this study represent (e.g., nominal, ordinal, interval, or ratio)? Why?

19. What kind of statistical test should be used to test the hypothesis (hint: think of what we have been doing in class lately)?

20. Which of the following are assumptions underlying the use of the paired *t* test?

    A. The variance of the population is known

    B. The sampling distribution is normal

    C. Data are interval or ratio

    D. All of the above

    E. A and B

    F. B and C

    G. A and C

21. A drug and alcohol researcher is interested in studying the effects of alcohol on learning ability of college seniors. She randomly assigns 10 students to an "alcohol group" and another 10 students to a control group. The students in

the alcohol group all receive 8 oz of alcohol prior to being tested. Then all of the students are run through a learning assessment and the number of errors is recorded. Assume the data below are normally distributed.

Control Group		Alcohol Group	
3		5	
2		3	
4	$\sum x = 27$	7	$\sum x = 38$
1	$\sum x^2 = 87$	2	$\sum x^2 = 166$
2	Mean = 2.7	4	Mean = 3.8
3	$s_{n-1} = 1.2517$	5	$s_{n-1} = 1.5492$
3		4	
1		3	
5		2	
3		3	

Perform the statistical test and state whether or not you can reject the null hypothesis.
The following homework questions should be answered with the online data set provided for this chapter via the textbook's website.

22. Produce a table of descriptive statistics using Microsoft Excel or SPSS.

23. Interpret the descriptive statistics produced by Excel. Do you meet the assumption of normality?

24. Do you meet the assumption of homogeneity of variance?

25. Analyze the data set using an independent t test and indicate if you have a statistically significant result. Explain how you evaluated the output.

PART III

ADDITIONAL HYPOTHESIS TESTS

CHAPTER 10
Analysis of Variance (ANOVA)

In Chapters 10 to 14, you will learn to analyze data from more complex experimental designs. In this chapter, we address the appropriate statistical test for situations in which you have three or more groups or conditions. This is a commonly used design in behavioral sciences because you can test for the effect of varying degrees of an independent variable. For example, you could test for the effects of caffeine on attention span by having high-caffeine, low-caffeine, and no-caffeine conditions. Alternatively, you could compare 4-year-olds, 8-year-olds, and 12-year-olds on their ability to use memory strategies. In this case, your independent variable is age, and you are using three levels of age in your experiment.

For now, assume your groups are independent of one another, so subjects only serve in one of the three conditions (also called a between-groups design). It is possible to have three or more groups participating in a within-groups (or repeated-measures) design, but we will discuss that situation in the next chapter.

WHY DO WE NEED THE ANOVA?

The analysis of variance (ANOVA) is similar to the independent t test discussed in Chapter 9, except that you have more than two groups. You may be wondering why you can't simply use the independent t test to analyze your data. Let's think about the number of tests you would need to perform to determine whether there were any significant differences between one or more of your groups or conditions. What if you had the following results from the experiment described above?

Low Caffeine	High Caffeine	No Caffeine (Control)
$\bar{X} = 20.88$	$\bar{X} = 13.15$	$\bar{X} = 30.29$

You would need to conduct *three* independent t tests to determine whether or not there were significant differences among these means. To be thorough, you would conduct the following comparisons:

Low caffeine versus high caffeine

Low caffeine versus no caffeine

High caffeine versus no caffeine

What's Wrong With Numerous Pairwise Comparisons?

When you perform multiple pairwise comparisons on the same data set, you run the risk of an inflated Type I error rate. Recall that we typically limit our Type I error rate by setting $\alpha = 0.05$. That means that we are willing to incorrectly reject the null hypothesis 5% of the time. However, when we make so many comparisons on the same data, we are making multiple "draws" from the same distribution. Our probability that we will be able to reject the null hypothesis for one of those comparisons increases, and thus our Type I error rate increases. One less desirable solution is to divide our α level by the number of comparisons we wish to make. This is called a Bonferroni adjustment. In the caffeine example, that would be $\frac{\alpha}{3}$, or $\frac{.05}{3} = 0.016666$. This is a very conservative approach to solving the Type I error rate for pairwise comparisons. If you had five groups/ conditions in your study, then you would have nine possible comparisons you could make among your means. If you applied this same "correction" to your α level, then you would be forced to use an α level of 0.0055 because $\frac{\alpha}{9}$, or $\frac{.05}{9}$ = 0.0055. Obviously, it would be difficult to obtain differences among your means large enough to demonstrate statistical significance. Fortunately, there is a better solution to the problem of controlling the Type I error rate without inflating the Type II error rate, as you are likely to do when you divide your α level by the number of comparisons you wish to make.

THE ANOVA SOLUTION

With the ANOVA, you perform one *overall* test to determine whether or not one or more of the conditions has different effects than the other conditions. Then, only if the overall test is statistically significant do you proceed to do pairwise comparisons to determine exactly which comparisons are different from one another.

Your alternative hypothesis (H_A) is that one or more conditions have an effect (one or more means differ from other means), and your null hypothesis (H_0) is that there is no effect (your means are equal). Mathematically, the null hypothesis would be represented in the following way:

$$\mu_1 = \mu_2 = \mu_3 = \ldots = \mu_k,$$

where k = number of groups/conditions.

You must assume that only the means are affected by your experimental manipulations and that the variance remains unaffected. Mathematically, this is written as

$$\sigma_1^2 = \sigma_2^2 = \sigma_3^2 = \ldots = \sigma_k^2.$$

Again, k represents the number of groups/conditions.

Specifically, we look at the variability between (or among) the group means. We assume this variability is due to our experimental manipulations and some natural inherent variation. We then compare it to the variability within the groups. Within-group variability should represent the inherent variation in the population. This can be more easily shown graphically. To show you how it works, we will first present a figure demonstrating this concept with the independent t test (two means), and then we can expand the demonstration to include three means.

s_W in this figure, as in the independent t test from the last chapter, denotes pooled estimate of variance from all samples. Note that the means in both examples are the same, but the variances are not the same. There is

Figure 10.1

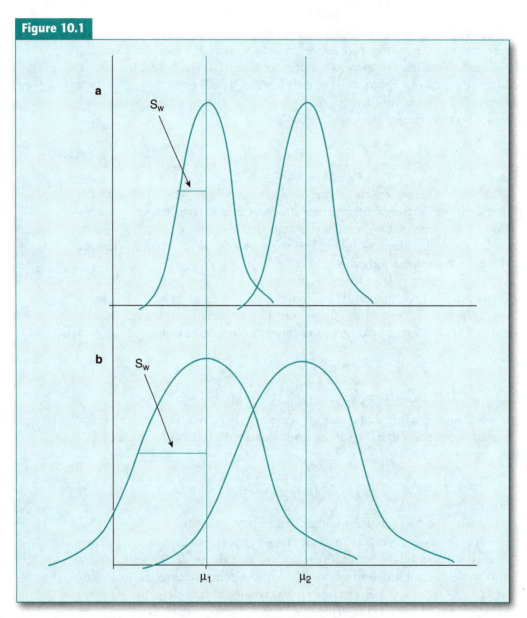

Comparing Means When There Is Low Versus High Within-Group Variability

more variability in Example b than you see in Example a, but the two groups in Example a have equal variances, as do the two groups in Example b. If both groups have the same variance (regardless of whether it is large or small), then you have met the assumption of homogeneity of variance. This means that you can now turn to analyzing the differences between the means, relative to the variability within the groups. This is essentially what you did when you performed the independent t test calculations in the last chapter, but you may not have thought about it in this way. If we take a moment to recall the formula for the independent t test, you'll see that we analyze both the differences between the means $(\bar{X}_1 - \bar{X}_2)$ as well as some measure of variability,

$$\sqrt{s_w^2(1/n_1 + 1/n_2)}.$$

Formula for an Independent t Test

$$t_{\text{obtained}} = \frac{\bar{X}_1 - \bar{X}_2}{\sqrt{s_w^2(1/n_1 + 1/n_2)}}.$$

The larger the numerator $(\bar{X}_1 - \bar{X}_2)$, or the larger the difference between your two sample means, and the smaller the denominator (your measure of variability), then the larger your t-obtained value will become. Of course, when you have a large t-obtained value, you are more likely to be able to reject the null hypothesis. Figure 10.1 shows two examples where the means are equal, but the variance is much higher in Example b compared to Example a. In this scenario, you can be sure that the t-obtained value for Example b will be much smaller than it will be for Example a. Thus, even when we were using the independent t test, the variability (or within-groups variance, s_w^2) played a role in determining the outcome of the analyses. When "within-group variance" is small relative to the difference between the means, we say there is a significant difference between our means. That continues to be true whether you have two independent (between-groups) samples or *more than* two samples, as is the case with the independent groups (between-groups) ANOVA.

In Figure 10.2, we again show two examples. This time, there are three groups with the same means, but the samples in Example a have smaller variances than the samples in Example b. Even though we now have three samples in each example (instead of two), the within-group variance is the same concept for ANOVA as it was for an independent t test. However, we can no longer just subtract the two means to determine the difference between them, so we now determine the variance among the means, or the "between-group variance, s_B^2." The between-group variance (s_B) is the average squared deviation of each group mean from the grand mean (overall mean of all subjects regardless of group). While the within-group variance (s_w^2) serves as an estimate of the true population

> s_w^2 : Within-group variance (reflects natural variation in the population, and thus it is an estimate of σ^2). Within-group variance could also be called "error" (E) because it is variability that we cannot control but occurs naturally.

Figure 10.2

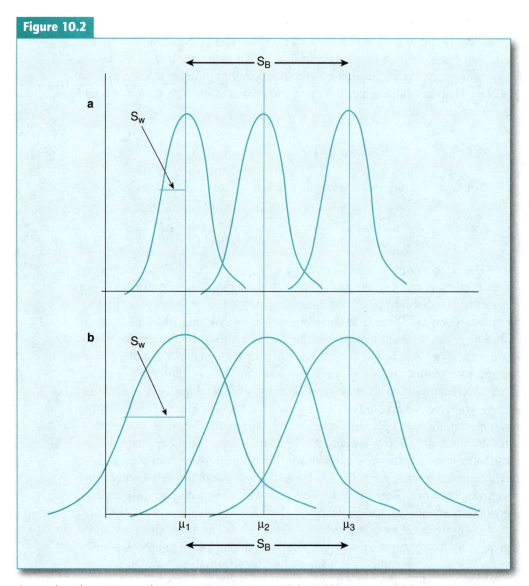

Comparing Three Means When You Have Low Versus High Within-Group Variability

s_B^2: Between-group variance (reflects natural variation in the population [E] and the effect of the independent variable, if any).

variance (σ^2), the between group variance $\left(s_B^2\right)$ includes that same natural variation as well as variability between groups that may be due to the effect of the experimental treatment. These are effects that shift the means away from one another.

Thus, each score is the result of starting with the grand mean (μ) and adding the variance from two sources: the treatment effect attributable to our independent variable (IV) and unexplained error (E).

Linear or Additive Equation

$$\text{Score} = \mu + \text{IV} + \text{E}.$$

There are multiple potential influences on the outcome for each score, and those influences can be defined with a linear equation. We will discuss this a bit more in Chapter 13. For now, it is important to focus on whether the effect of our independent variable (IV) is enough to demonstrate a statistically significant difference among our groups. To test for that difference, we will calculate an *F*-obtained value, where *F* is the symbol for the ANOVA statistic, and it is expressed with the following equation:

$$F = \frac{s_B^2}{s_W^2} = \frac{\text{estimated } \sigma^2 \text{ and treatment}}{\text{estimated } \sigma^2}.$$

The natural variability (estimated σ^2) cancels itself out in the numerator and denominator:

$$F = \frac{s_B^2}{s_W^2} = \frac{\text{estimated } \sigma^2 \text{ and } \boxed{\text{treatment}}}{\text{estimated } \sigma^2},$$

so what we have left is the effect of our treatment.

Figure 10.3 demonstrates two scenarios where the s_w^2 is the same for all of the groups and for both examples, but the means differ significantly in Example a. In this situation, you would expect a larger s_B^2 in Example a because your measure of the differences among the means will be larger for Example a than it will be for Example b. Thus, you are more likely to calculate an *F*-obtained value that is greater than 1.00 in Example a than you are for Example b.

Properties of Your *F*-Obtained Value

Our *F*-obtained value should be a measure of our experimental treatment, or the effect of our experiment. If there is no effect, then *F*-obtained will equal 1.00 (approximately). The reason that *F*-obtained will equal 1.00 when there is no treatment effect is because you would essentially be dividing your estimated variance by itself, and any number divided by itself is equal to 1.00. Also, *F* can never have a negative value since it is the ratio of two squared values, and squared values are always positive.

Figure 10.3

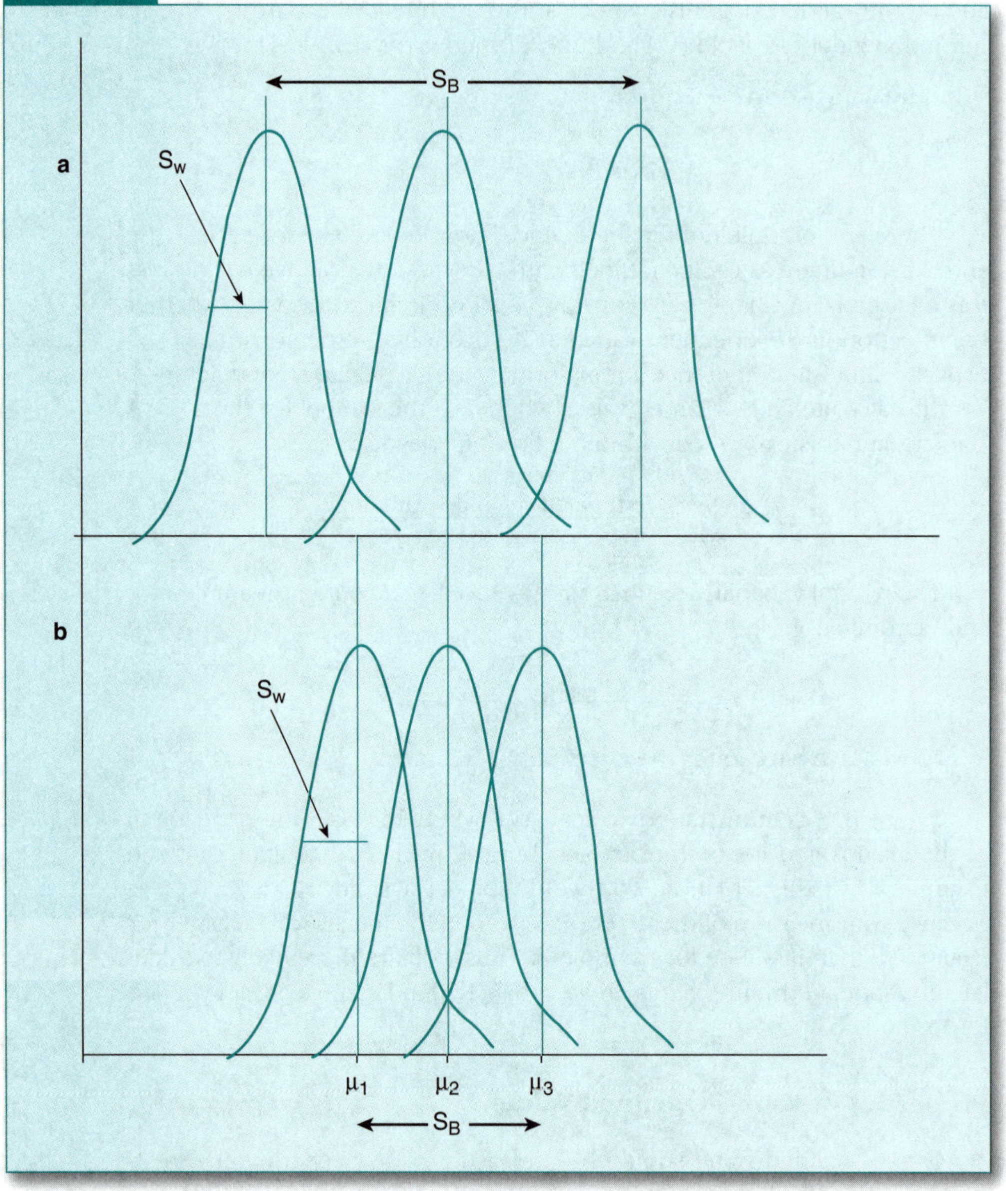

Comparing Different Means When Within-Group Variability Is the Same

Of course, F-obtained will rarely equal exactly 1.00 since our two esti-mates of variance are just that, estimates. Each estimate will vary slightly due to random effects, but F-obtained values near 1.00 will never be significant because they will suggest little or no effect of our experimental treatment.

However, if we divide the between-groups variance by the within-groups variance and get a number greater than 1.00, then we have a statistic that measures our treatment effect. If the value is large enough to be unlikely to occur by chance, then we can say that our treatment (our independent variable) had a significant effect on our dependent variable. Thus, a larger *F*-obtained value means a greater effect of the independent variable.

Calculating an *F*-Obtained Value

Recall that variances are calculated dividing the sum of squares (SS) by the degrees of freedom (*df*). Thus, the between- and within-groups variance calculations involve the appropriate SS and *df*.

$$s_B^2 = \frac{SS_B}{df_B}$$

$df_B = k - 1$, where k = number of groups/conditions

$$s_w^2 = \frac{SS_W}{df_W}$$

$df_W = N_T - k$, where N_T = total number of subjects in all conditions

$$F_{\text{obtained}} = \frac{s_B^2}{s_w^2}$$

Calculating the SS values is done in the usual way by adding and squaring the scores, but it is trickier than it was for the *t* tests because there are more than two groups, and you must calculate the $\sum X$ for each group and for all of the subjects in the experiment. Nevertheless, it is conceptually similar to the SS calculations that you have performed in past chapters. You can refer to the sidebars for more information on calculating the SS manually.

FYI: FORMULAS

$$SS_W = \overset{\substack{all \\ scores}}{\sum} X^2 - \left[\frac{(\Sigma X_1)^2}{n_1} + \frac{(\Sigma X_2)^2}{n_2} + \frac{(\Sigma X_3)^2}{n_3} + \ldots + \frac{(\Sigma X_k)^2}{n_k} \right]$$

$$SS_B = \left[\frac{(\Sigma X_1)^2}{n_1} + \frac{(\Sigma X_2)^2}{n_2} + \frac{(\Sigma X_3)^2}{n_3} + \ldots + \frac{(\Sigma X_k)^2}{n_k} \right] - \frac{\left(\overset{\substack{all \\ scores}}{\sum} X \right)^2}{N}$$

(Continued)

(Continued)

where

$$\sum_{\text{all scores}} X^2 = \text{sum of all of the squared scores (regardless of group/condition)}$$

$$\sum_{\text{all scores}} X = \text{sum of all of the scores (regardless of group/condition)}$$

$\sum X_1$ = sum of the scores for Group 1

n_1 = number of scores in Group 1

$\sum X_2$ = sum of the scores for Group 2

n_2 = number of scores in Group 2

$\sum X_3$ = sum of the scores for Group 3

n_3 = number of scores in Group 3

$\sum X_k$ = sum of the scores for Group k (last group)

n_k = number of scores in Group k

N = total number of scores (regardless of group/condition) or sum of $n_1 + n_2 + n_3 + \ldots + n_K$

We have chosen to give you the SS values to save time and because you can understand how F-obtained is influenced by variability without manually calculating the SS for each group and for the entire experiment. Keep in mind that the SS_B and the SS_W can be added together to create the SS total (SS_T). Thus, if you know the SS_T and one of the other SS, you can determine the other value by subtraction.

$$SS_T = SS_B + SS_W$$

Table 10.1

Source	SS	df	s^2 (MS)	F Obtained
Between groups				
Within groups				
Total				

ANOVA Summary Table (Sample)

The formulas listed above can be used to fill out the ANOVA summary table (see Table 10.1). ANOVA results are traditionally presented in this format. By dividing the SS by the appropriate *df*, you can fill in the variance (s^2) column. Since variance is the average squared deviations around the mean, in ANOVA, variance is often referred to as "mean square" or MS. Dividing the variance between groups by the variance within groups gives you the *F*-obtained value. Given these "formulas," you can even complete an ANOVA summary table that is only partially complete if you know a few values and the mathematical relationships between them. For example, you could be given the *df* and variance values but not the SS values. You could fill in the SS values by multiplying *df* and variance separately for between and within groups. Since $\frac{SS}{df} = s^2$, then $s^2 * df = SS$.

CHARACTERISTICS OF THE *F* DISTRIBUTION

The *F* distribution, like the *t* distribution, is based on a family of curves that varies with the degrees of freedom. In this case, we have different degrees of freedom for the numerator (df_B) and the denominator (df_W) of the *F* formula, and there is a different curve for every possible combination of df_B and df_W. As you can imagine, there are unlimited combinations of these two values, and thus we will rely on critical values to indicate the values that *F*-obtained must equal, or be more extreme than, to achieve statistical significance. Because *F*-obtained always yields a positive number, it is not too surprising that the *F* distribution is also positive and positively skewed. In addition, all tests of ANOVA are nondirectional. The *F* distribution table in the Appendix (Table C) contains the *F*-critical values that are most likely to be used in this course.

Complete Example

A researcher measures the hesitation time in three samples of participants in a study of quick decision making. The three groups represent three ethnic backgrounds: Caucasian (Group A), African American (B), and Hispanic (C), and the study asks if there is a difference in hesitation time among the three groups. Assume that the data are normally distributed and the variances are homogeneous. The dependent variable is hesitation time measured in seconds. Let $\alpha = 0.05$.

Hesitation Time (s)		
Group A	**Group B**	**Group C**
2.5	2.8	3.8
3.1	3.2	3.2
2.2	3.0	4.0
2.6	2.7	3.7
2.2	2.3	3.7

Null hypothesis: There is no difference in the hesitation time among the three groups or the differences that exist are due to chance.

Alternative hypothesis: There is a difference in the hesitation time of one or more groups from one or more of the other groups.

 Step 1: Compute the probability that each of the sample means comes from the null hypothesis population where $\mu_1 = \mu_2 = \mu_3$.

Calculate the means and the degrees of freedom for each group:

$$\sum X_A = 12.6 \qquad \sum X_B = 14 \qquad \sum X_C = 18.4.$$

$$\bar{X}_A = \frac{12.6}{5} = 2.52 \qquad \bar{X}_B = \frac{14}{5} = 2.80 \qquad \bar{X}_C = \frac{18.4}{5} = 3.68.$$

$df_B = k - 1$, where k = number of groups/conditions, so $df_B = 3 - 1 = 2$.

$df_W = N_T - k$, where N_T = total number of participants in all conditions, so $df_W = 15 - 3 = 12$.

Note: You can check your work on the degrees of freedom by calculating the df_T.

$$df_T = N_T - 1 = df_B + df_W, \text{ so } df_T = 15 - 1 = 2 + 12 = 14.$$

Apply the s^2 and F-obtained formulas to complete the ANOVA summary table (Table 10.2).

Numbers in bold were filled in by solving the formulas while df (not bolded) were calculated in the previous step. Calculations for the SS are available in the sidebar.

Table 10.2

ANOVA Summary Table				
Source	SS	df	s^2 (MS)	F Obtained
Between groups	3.664	2	**1.832**	**16.212**
Within groups	1.356	12	**0.113**	
Total	**5.020**	14		

Hesitation Time Example

$$SS_T = SS_B + SS_W$$

$$SS_T = 3.664 + 1.356 = 5.020$$

$$s_B^2 = \frac{SS_B}{df_B} = \frac{3.664}{2} = 1.832$$

$$s_w^2 = \frac{SS_W}{df_W} = \frac{1.356}{12} = 0.113$$

$$F_{\text{obtained}} = \frac{s_B^2}{s_W^2} = \frac{1.832}{0.113} = 16.212$$

FYI: SS FOR SPECIES

$$\sum_{\substack{all \\ scores}} X^2 = 140.02$$

$$\sum_{\substack{all \\ scores}} X = 45$$

$$\sum X_1 = 12.6$$

$$\sum X_2 = 14$$

$$\sum X_3 = 18.4$$

$$n_1 = 5$$

$$n_2 = 5$$

$$n_3 = 5$$

$$N = 15$$

(Continued)

(Continued)

$$SS_W = \sum_{}^{\substack{all \\ scores}} X^2 - \left[\frac{(\Sigma X_1)^2}{n_1} + \frac{(\Sigma X_2)^2}{n_2} + \frac{(\Sigma X_3)^2}{n_3} + \dots + \frac{(\Sigma X_k)^2}{n_k} \right]$$

$$SS_W = 140.02 - \left[\frac{(12.6)^2}{5} + \frac{(14)^2}{5} + \frac{(18.4)^2}{5} \right]$$

$$SS_W = 140.02 - \left[\frac{158.76}{5} + \frac{196}{5} + \frac{338.56}{5} \right]$$

$$SS_W = 140.02 - [31.752 + 39.2 + 67.712]$$

$$SS_W = 140.02 - 138.664 = 1.356$$

$$SS_B = \left[\frac{(\Sigma X_1)^2}{n_1} + \frac{(\Sigma X_2)^2}{n_2} + \frac{(\Sigma X_3)^2}{n_3} + \dots + \frac{(\Sigma X_k)^2}{n_k} \right] - \frac{\left(\sum_{}^{\substack{all \\ scores}} X \right)^2}{N}$$

$$SS_B = \left[\frac{(12.6)^2}{5} + \frac{(14)^2}{5} + \frac{(18.4)^2}{5} \right] - \frac{(45)^2}{15}$$

$$SS_B = [31.752 + 39.2 + 67.712] - \frac{2025}{15}$$

$$SS_B = 138.664 - 135 = 3.664$$

★ **Step 2:** Evaluate the probability of obtaining this score due to chance.

Evaluate the F-obtained value based on alpha (α) = 0.05. To evaluate your F-obtained value, you must use the F distribution table in the Appendix (Table C), and to do so, you need an alpha level (0.05) and degrees of freedom (df_B and df_W). The degrees of freedom from the numerator of the F-obtained equation are the df_B, while the degrees of freedom from the denominator are df_W. Thus, the df for this problem are 2 and 12. Compare the F-critical value with your F-obtained value, and reject the null hypothesis when your F-obtained value is equal to or greater than your F-critical value. When α = 0.05 and the degrees of freedom are equal to 2 and 12, you should use 3.88 as your F-critical value.

16.212 ≥ 3.88, so we *reject* the null hypothesis.

How should we interpret these data in light of the hesitation time among these three groups?

These results suggest that you would expect to obtain these differences less than 5% of the time, if ethnic group membership had no effect. Thus, it is not very likely that these hesitation length differences all come from the same null hypothesis population. Of course, as always, there is some finite probability that you could get a difference this large among three means that is purely due to chance, but that probability is less than 5%.

In addition, we can conclude that one or more of these three means is significantly different from one or more of the other means. We also know from our calculations of the means above that Group C has the longest hesitation time and that Group A has the shortest hesitation time, so it is likely that we have a significant difference between these two groups, but what about B versus C or A versus B?

Determining the statistical significance of these pairwise comparisons can be done using "multiple comparisons" once you know that your overall test (your ANOVA) has demonstrated statistical significance. We will turn to this concept in the next section of this chapter, but we will first take a look at the Excel and SPSS output for these data that we have analyzed by hand.

Results if you use Microsoft Excel to calculate the ANOVA:

Anova: Single Factor						
Summary						
Groups	**Count**		**Sum**	**Average**		**Variance**
Group A	5		12.6	2.52		0.137
Group B	5		14	2.8		0.115
Group C	5		18.4	3.68		0.087
ANOVA						
Source of Variation	**SS**	**df**	**MS**	**F**	**P-value**	**F crit**
Between Groups	3.664	2	1.832	16.21239	0.000388	**3.88529**
Within Groups	1.356	12	0.113			
Total	**5.02**	14				

This is the output that you get from Excel when you type in the data for the three ethnic groups. Note that Excel calls this test an "ANOVA: Single Factor." This wording is slightly different from what we have been using, but it is describing the same analysis; this will be even clearer after we have discussed other types of ANOVA analyses in the next chapter. We have bolded the numbers that are comparable to the numbers that we just manually calculated or looked up in the table. Note that the F-obtained value (Excel calls it "F") is identical to ours, except they have carried more digits (16.21239). The "F Crit" is also the same (3.88529), but again they have carried more digits. The degrees of freedom are the same (2 and 12). In addition, they have provided you with the exact probability ("P-value") of obtaining an F-obtained value ("F") of 16.21239, which is 0.000388. Since 0.000388 < 0.05, we clearly *reject* our null hypothesis. These results suggest that exactly 0.0388% of the time, you would obtain this difference in hesitation times if ethnic group membership had no effect. Again, knowing the exact probability (not just the F-critical values) for each F-obtained value, along with computation speed, is one of the major advantages of using a computer to calculate your analyses.

Results if you use SPSS to calculate the ANOVA (or *F* test):

	Sum of Squares	df	Mean Square	F	Sig.
Between Groups	3.664	**2**	**1.832**	**16.212**	.000
Within Groups	1.356	**12**	**.113**		
Total	**5.020**	**14**			

ANOVA

HESITATION

We have placed the numbers that are comparable to our manual calculations in bold. Once again, you see that the exact probability (SPSS calls it "Sig.") is less than our alpha level of 0.05, and thus we must assume that these results would be *unlikely* simply due to chance.

How would these results be reported in a scientific journal article?

If the journal required American Psychological Association (APA) format, the results would be reported in a format something like this:

There is at least one significant difference in the hesitation time among the three ethnic groups, $F(2, 12) = 16.21, p < .01$.

This formal sentence includes the dependent variable (hesitation time), the independent variable (three groups), as well as a statement about statistical significance, the symbol of the test (F), the degrees of freedom between groups (2), the degrees of freedom within groups (12), the statistical value (16.21), and the probability of obtaining this result simply due to chance ($< .01$).

Another Complete Example

Teachers in the K–12 grades all too frequently "burn out" and leave the teaching profession. A researcher suggests that a combination (one or more) of four assessments of stress and professional efficacy might provide the option to anticipate such problems and take action to prevent losing good teachers. The immediate question is whether scores from the four assessments are all the same across teachers or whether there is a difference in scores for the assessments. A group of 40 teachers are randomly assigned to four equal groups, and each group is given one of the assessments. Is there a significant difference in mean scores for the four groups? Assume that the data are normally distributed and the variances are homogeneous. The dependent variable is assessment score. Let $\alpha = 0.05$.

Group A	Group B	Group C	Group D
16.3	18.3	13.6	16.7
23.2	21.3	14.6	18.2
21.0	19.5	19.0	18.6
19.0	18.6	21.0	19.5
21.9	21.4	14.9	12.6
16.4	18.9	21.9	13.7
14.9	12.6	21.9	21.4
21.9	13.7	16.4	18.9
13.6	16.7	16.3	18.3
14.6	18.2	23.2	21.3

Assessment Scores

Null hypothesis: There is no difference in the assessment scores among the four groups or types of assessments, or the differences are due to chance.

Alternative hypothesis: There is a difference in the mean assessment scores obtained from one or more of the assessments.

 Step 1: Compute the probability that each of the sample means comes from the null hypothesis population where $\mu_1 = \mu_2 = \mu_3 = \mu_4$.

Calculate the means and the degrees of freedom for each group:

$$\sum X_A = 182.8 \qquad \sum X_B = 179.2 \qquad \sum X_C = 176.7 \qquad \sum X_D = 185.3.$$

$$\bar{X}_A = 18.28 \qquad \bar{X}_B = 17.92 \qquad \bar{X}_C = 17.67 \qquad \bar{X}_D = 18.53.$$

$df_B = k - 1$, where k = number of groups/conditions, so $df_B = 4 - 1 = 3$.

$df_W = N_T - k$, where N_T = total number of subjects in all conditions, so $df_W = 40 - 4 = 36$.

Note: You can check your work on the degrees of freedom by calculating the df_T.

$$df_T = N_T - 1 = df_B + df_W, \text{ so } df_T = 40 - 1 = 3 + 36 = 39.$$

Apply the s^2 and F-obtained formulas to complete the ANOVA summary table (Table 10.3).

Table 10.3

ANOVA Summary Table				
Source	**SS**	**Df**	**s² (MS)**	**F Obtained**
Between Groups	4.35	3	**1.45**	**0.14**
Within Groups	372.81	36	**10.36**	
Total	**377.16**	39		

Burnout Assessment Example

Numbers in bold were filled in by solving the formulas, while *df* (not bolded) were calculated in the previous step.

$$SS_T = SS_B + SS_W$$

$$SS_T = 4.35 + 372.81 = 377.16$$

$$s_B^2 = \frac{SS_B}{df_B} = \frac{4.35}{3} = 1.45$$

$$s_w^2 = \frac{SS_W}{df_W} = \frac{372.81}{36} = 10.36$$

$$F_{obtained} = \frac{s_B^2}{s_W^2} = \frac{1.45}{10.36} = 0.14$$

 Step 2: Evaluate the probability of obtaining this score due to chance.

Evaluate the F-obtained value based on alpha (α) = 0.05. To evaluate your F-obtained value, you must use the F distribution table in the Appendix (Table C), and to do so, you need an alpha level (0.05) and degrees of freedom (df_B and df_W). The degrees of freedom from the numerator of the F-obtained equation are the df_B, while the degrees of freedom from the denominator are df_W. Thus, the df for this problem are 3 and 36. Compare the F-critical value with your F-obtained value, and reject the null hypothesis when your F-obtained value is equal to or greater than your F-critical value. When $\alpha = 0.05$ and the degrees of freedom are equal to 3 and 36, you should use 2.86 as your F-critical value.

0.14 < 2.86, so we *fail to reject* the null hypothesis.

How should we interpret these data in light of the differences in assessment outcomes?

These results suggest that you would expect to obtain these differences in assessment outcomes more than 5% of the time, if there was no difference in the assessments. Thus, it is very likely that these assessment outcomes all come from the same normal null hypothesis population. Of course, as always, there is some finite probability that there is a difference in the outcomes of one or more of these assessments that you are not detecting with this analysis.

Unlike the previous example, there is no need to determine the statistical significance of the pairwise comparisons among these four means since the overall F-test has failed to reject the null hypothesis. If the overall test does not detect a difference among your means, you *do not* continue to the pairwise comparisons of the means: You are done with the analysis. Let's take a look at the Excel and SPSS output for these data that we have analyzed by hand.

Results if you use Microsoft Excel to calculate the ANOVA:

Summary				
Groups	**Count**	**Sum**	**Average**	**Variance**
Group A	**10**	**182.8**	**18.28**	12.517
Group B	**10**	**179.2**	**17.92**	8.364
Group C	**10**	**176.7**	**17.67**	10.142
Group D	**10**	**185.3**	**18.53**	10.400

ANOVA						
Source of Variation	**SS**	**Df**	**MS**	**F**	**P-value**	**F crit**
Between Groups	4.346	**3**	**1.449**	**0.140**	0.935	**2.866**
Within Groups	372.814	**36**	**10.356**			
Total	**377.160**	**39**				

This is the output that you get from Excel when you type in the data for the four assessment groups. We have bolded the numbers that are comparable to the numbers that we just manually calculated or looked up in the table. Note that these results are the same as our hand-calculated example. In addition, they have provided you with the exact probability ("P-value") of obtaining an F-obtained value ("F") of 0.14, which is 0.935. Since $0.935 > 0.05$, we clearly *fail to reject* our null hypothesis. These results suggest that exactly 93.5% of the time, you would obtain this difference in assessment scores if the assessments were not different.

Results if you use SPSS to calculate the ANOVA (or F test):

	Sum of Squares	Df	Mean Square	F	Sig.
Between Groups	4.346	**3**	**1.449**	**.140**	.935
Within Groups	372.814	**36**	**10.356**		
Total	**377.160**	**39**			

ANOVA

SCORES

We have placed the numbers that are comparable to our manual calculations in bold. Once again, you see that the exact probability (SPSS calls it "Sig.") is greater than our alpha level of 0.05, and thus we must assume that these results would be *likely* due to chance and therefore do not reflect real differences among the assessments.

How would these results be reported in a scientific journal article?

If the journal required APA format, the results would be reported in a format something like this:

> There is no significant difference in the assessment scores among the four groups or types of assessments, $F(3, 36) = .14, p = .93$.

This formal sentence includes the dependent variable (assessment scores), the independent variable (four groups), as well as a statement about statistical significance, the symbol of the test (F), the degrees of freedom between groups (3), the degrees of freedom within groups (36), the statistical value (.14), and the probability of obtaining this result simply due to chance (.93).

When is it appropriate to use an ANOVA test?

1. When you have more than two samples and a between-groups design

2. When the sampling distribution is normally distributed. Again, this is satisfied when
   a. The sample size is greater or equal to 30 in each group ($n \geq 30$) or
   b. The null hypothesis population is known to be normally distributed.

3. When the dependent variable is on an interval or ratio scale

4. When the variances of the groups are the same or are homogeneous. The homogeneity of variance assumption (HOV) requires that the variances of the samples are equal or, in practical terms, not significantly different from one another. We will use a simple rule of thumb to decide whether the variances are similar enough to be considered homogeneous: if the $\frac{\text{largest variance}}{\text{smallest variance}} \leq 4$, then the HOV assumption is met (okay) and you can proceed with ANOVA. If not, you cannot proceed with the ANOVA (later, we'll explain what to do instead).

If these assumptions are not met, it may be possible to transform the data to create homogeneous variances or distributions that more closely resemble a normal distribution, but a common alternative is to change to a nonparametric test such as the Kruskal-Wallis test. We will describe this test in detail in Chapter 14.

MULTIPLE COMPARISONS

If you are fortunate enough to obtain statistical significance with your ANOVA (*F* test), then you will need to do multiple comparisons to determine which of your means is significantly different from which of your other means. There are two major categories of comparisons, planned and unplanned. We will overview how to do the planned comparisons first and then move to the more common unplanned or post hoc comparisons.

Planned (A Priori) Comparisons

Planned comparisons are performed whenever you have preplanned a *limited* number of comparisons prior to collecting your data. Planned comparisons are also called "a priori" comparisons because *a priori* is a Latin term that means "from what goes before; from cause to effect." Because you have preplanned these comparisons, typically based on prior data and theory, and you do not plan to do *all possible* comparisons, you are not required to make a correction for your alpha (α) level. In this situation, it is appropriate to conduct independent *t* tests to determine if your planned comparisons have means that are significantly different from one another. Fortunately, this calculation is simplified based on the calculations you have already done for the ANOVA summary table. You are able to obtain the s_w^2 (a better estimate of variance than just taking the average of two estimates) directly from the ANOVA summary table to insert into your independent *t* test formula:

$$t_{\text{obtained}} = \frac{\bar{X}_1 - \bar{X}_2}{\sqrt{s_w^2 (1/n_1 + 1/n_2)}}.$$

Ethnic Group Example

If we revisit the hesitation time analysis that we performed as an ANOVA, we can now determine whether Group A versus Group C are significantly different and whether Group B versus Group C are significantly different from one another by performing two independent *t* tests. Note that we are *not* performing a *t* test on all possible comparisons because we are not comparing A to B. We would have had to preplan the two comparisons that we are performing if this had been a real study. For your convenience, we have repeated the ANOVA summary table from SPSS to refer to as we demonstrate the calculations.

$$t_{obtained} = \frac{\bar{X}_1 - \bar{X}_2}{\sqrt{s_w^2(1/n_1 + 1/n_2)}}.$$

ANOVA: Single Factor				
Summary				
Groups	**Count**	**Sum**	**Average**	**Variance**
Group A	5	12.6	2.52	0.137
Group B	5	14	2.8	0.115
Group C	5	18.4	3.68	0.087
ANOVA				
Source of Variation	**SS**	**df**	**MS**	**F**	**P-value**	**F crit**
Between Groups	3.664	2	1.832	16.21239	0.000388	3.88529
Within Groups	1.356	12	0.113			
Total	5.02	14				

 Step 1: Calculate the *t*-obtained values for Group A versus Group C and Group B versus Group C.

$$t_{obtained} = \frac{2.52 - 3.68}{\sqrt{0.113(1/5 + 1/5)}} = \frac{-1.16}{\sqrt{0.113(.40)}} = \frac{-1.116}{0.2126} = -5.46 \text{ (A vs. C)}.$$

$$t_{obtained} = \frac{2.80 - 3.68}{\sqrt{0.113(1/5 + 1/5)}} = \frac{-0.88}{\sqrt{0.113(.40)}} = \frac{-0.88}{0.2126} = -4.14 \text{ (B vs. C)}.$$

 Step 2: Evaluate the probability of obtaining this score due to chance.

Evaluate the *t*-obtained value based on alpha (α) = 0.05 and a nondirectional hypothesis. The type of hypothesis used here must match the (always) nondirectional hypothesis of the main ANOVA test. To evaluate your *t*-obtained value, you must use the *t* distribution (Table B in the Appendix). To determine your *t*-critical value, you need to know your alpha level (0.05), the number of tails you are evaluating (always two in this case), and your degrees of freedom (*df*). We use the degrees of freedom associated with the s_w^2, or the df_w, to evaluate the *t*-obtained value. Thus, the *df* for this problem are 12. Compare the *t*-critical value with your *t*-obtained value. When $\alpha = 0.05$, your degrees of freedom are equal to 12, and your hypothesis is two-tailed, you should use 2.179 as your *t*-critical value.

$-5.46 \geq |2.179|$, so we *reject* the null hypothesis that there is no difference between Groups A and C.

$-4.14 \geq |2.179|$, so we *reject* the null hypothesis that there is no difference between Groups B and C.

How should we interpret these data in light of the effect of ethnic background on hesitation time?

These results suggest that *less than* 5% of the time, you would obtain these differences in hesitation time if there was no effect of ethnic group membership. Thus, it is not very likely that these hesitation times for Group A versus Group C and Group B versus Group C come from the normal null hypothesis population of mean differences. However, there is a chance that you could get a difference this large between two means that is purely due to chance, but that chance is less than 5%.

How would these results be reported in a scientific journal article?

If the journal required APA format, the results would be reported in a format something like this:

> There is a significant difference in the hesitation times of Caucasians and Hispanics, $t(12) = -5.46, p < .05$. The hesitation time was longer in Hispanics. There was a significant difference in the hesitation times of African Americans and Hispanics, $t(12) = -4.14, p < .05$. Again, the hesitation time was longer in Hispanics.

These formal sentences include the dependent variable (hesitation time), the independent variable (ethnic group), as well as a statement about statistical significance, the symbol of the test (*t*), the degrees of freedom within groups (12), the statistical value (–5.46 and –4.14), and the probability of obtaining these results simply due to chance ($< .05$).

Unplanned (A Posteriori) or Post Hoc Comparisons

Unplanned comparisons are often called post hoc or "a posteriori" comparisons. *A posteriori* is a Latin phrase that means "from what follows; from effect to cause." *Post hoc* is an abbreviation for a Latin phrase *post hoc ergo propter hoc,* which means "after this therefore because of this." This phrase describes the fallacy that can occur when you assume that because one thing follows another that the one thing was caused by the other. Of course, not planning a limited number of comparisons prior to the start of the study also means that you must perform comparisons that limit your Type I error rate. While this is less desirable than planned comparisons, it is often necessary when there is no strong prior knowledge regarding which groups are likely to differ.

Many different techniques limit your Type I error rate. Some of these techniques are considered liberal (meaning that it easier to reject the null hypothesis), while others are considered conservative (harder to reject the null hypothesis). Because there are many different post hoc tests, and most computer statistical packages will do a variety of them, we will focus on one to calculate manually to demonstrate how the test can limit your probability of making a Type I error. The test that we will show is Tukey's honestly significant difference (HSD) test. This test maintains your alpha level for the full set of comparisons, and thus it controls your Type I error rate at the level of the experiment rather than for each comparison. This is a conservative test in that it more strictly controls the probability of making a Type I error. The number of means being compared is determined by comparing each mean with all of the other means. For example, what if you had four means and you wanted to make every possible comparison between them? If you had four means, then you would need to perform six post hoc comparisons, which would be the following:

1 vs. 2 AND 1 vs. 3 AND 1 vs. 4 AND 2 vs. 3 AND 2 vs. 4 AND 3 vs. 4.

Q Distribution

The Tukey's HSD test, like the statistical tests you have already learned, has its own sampling distribution that is used to evaluate whether our obtained values are likely due to chance alone. The *Q* distribution was created by randomly taking samples of equal n and calculating the difference between the most extreme means. This created a distribution that tells us how often you can get two sample means that differ by the amount calculated with your Tukey HSD test. As we have done before, we will reject the null hypothesis that suggests that there is no difference between the means only if the calculated *Q* statistic is more than a critical value of *Q*.

Formula for Calculating a Tukey HSD Comparison (*Q* Obtained)

$$Q \text{ obtained} = \frac{\bar{X}_i - \bar{X}_j}{\sqrt{\frac{s_w^2}{n}}}$$

where $\bar{X}_i$ = larger of the two means and $\bar{X}_j$ = smaller of the two means.

Note: We limit our examples and exercises to situations where the sample sizes are equal. It is possible to perform this test on unequal samples sizes: Refer to a more advanced statistics text for instructions.

Complete Example

Let's return to our hesitation time example and analyze all possible comparisons of the three means using the Tukey HSD approach. Evaluate the results based on $\alpha = 0.05$.

★ **Step 1:** List the group means.

$$\bar{X}_A = 2.52 \qquad \bar{X}_B = 2.80 \qquad \bar{X}_C = 3.68$$

★ **Step 2:** Calculate Q-obtained values for all possible comparisons.

$$Q \text{ obtained} = \frac{\bar{X}_i - \bar{X}_j}{\sqrt{\frac{s_w^2}{n}}}$$

A vs. B: Q obtained $= \dfrac{2.80 - 2.52}{\sqrt{\frac{0.113}{5}}} = 1.863$

A vs. C: Q obtained $= \dfrac{3.68 - 2.52}{\sqrt{\frac{0.113}{5}}} = 7.718$

B vs. C: Q obtained $= \dfrac{3.68 - 2.8}{\sqrt{\frac{0.113}{5}}} = 5.855$

where Group 1 = A, Group 2 = B, and Group 3 = C.

★ **Step 3:** Evaluate Q obtained based on Q-critical values.

Evaluate the Q-obtained values based on alpha $(\alpha) = 0.05$, the number of means that you are comparing (three groups, so $k = 3$), and the df_w. To evaluate your Q-obtained values, you must use the Q distribution (Table D) in the Appendix). When $\alpha = 0.05$, your degrees of freedom are equal to 12, and $k = 3$, you should use 3.77 as your critical value.

1.863 < 3.77, so we *fail to reject* the null hypothesis that there is no difference between Groups A and B.

7.718 ≥ 3.77, so we *reject* the null hypothesis that there is no difference between Groups A and C.

5.855 ≥ 3.77, so we *reject* the null hypothesis that there is no difference between Groups B and C.

How should we interpret these data in light of the effect of ethnic background on hesitation time?

These results suggest that for Group A versus Group C and Group B versus Group C, there is less than a 5% chance that you would obtain these differences in hesitation time if there was no effect of ethnic group membership. Thus, it is not very likely that these hesitation times for Group A versus Group C and Group B versus Group C come from the normal null hypothesis population of mean differences. However, there is a chance that you could get a difference this large between two means that is purely due to chance, but that chance is less than 5%.

There appears to be no (detectable) difference in hesitation times between Group A and Group B, and we must conclude that these hesitation times come from the same underlying population: There is no difference in hesitation time between Caucasians and African Americans.

Results if you use SPSS to calculate the Tukey HSD:

	(I) GROUP	(J) GROUP	Mean Difference (I-J)	Std. Error	Sig.	95% Confidence Interval Lower Bound	Upper Bound
Tukey HSD	1.00	2.00	−.2800	.2126	**.413**	−.8472	.2872
		3.00	−1.1600	.2126	**.000**	−1.7272	−.5928
	2.00	1.00	.2800	.2126	.413	−.2872	.8472
		3.00	−.8800	.2126	**.004**	−1.4472	−.3128
	3.00	1.00	1.1600	.2126	.000	.5928	1.7272
		2.00	.8800	.2126	.004	.3128	1.4472

Tukey HSD

Multiple Comparisons

Dependent Variable: HESITATION TIMES

This is the output you get from SPSS when you type in the data for the three groups. Note that SPSS calls this test a "Tukey or Tukey HSD." The numbers in bold indicate the calculated probabilities (.413 for Group 1 vs. Group 2; .000 for Group 1 vs. Group 3; .004 for Group 2 vs. Group 3; note that SPSS labels the groups as 1, 2, 3 rather than as A, B, and C). Note that SPSS is somewhat redundant in that it calculates the probability (or "Sig.") of 1 versus 2 as well as 2 versus 1. However, the conclusion that the means for Samples 1 and 2 are not significantly different, but the mean for Sample 3 is significantly different from the means for Samples 1 and 2, is the same conclusion that we obtained in doing the problem by hand.

How would these results be reported in a scientific journal article?

If the journal required APA format, the results would be reported in a format something like this:

> There is no significant difference in the hesitation times of Caucasians and African Americans, $Q(12) = 1.86, p > .05$. There is a significant difference in the hesitation times of Caucasians and Hispanics, $Q(12) = 7.72, p < .05$. There is a significant difference in the hesitation times of African Americans and Hispanics, $Q(12) = 5.86, p < .05$. Hesitation time was longest in the Hispanic group.

These formal sentences include the dependent variable (hesitation time), the independent variable (ethnic background), as well as a statement about statistical significance, the symbol of the test (Q), the degrees of freedom within groups (12), the statistical value (1.86, 7.72, and 5.86), and the probability of obtaining these results simply due to chance ($> .05, < .05, < .05$).

EFFECT SIZES AND POWER

Effect size is a standardized measure of the difference between two (or more) group means; it is the difference in means divided by the shared standard deviation of the two or more groups. For ANOVA, you can estimate the effect size by using Cohen's f^2:

$$\text{Cohen's } f^2 = \frac{R^2}{1 - R^2}.$$

Note: R^2 can be obtained from your statistical outputs, particularly if you set the output for the "long" or detailed version.

If you do not know your effect size, you can still proceed by using software and the values from your statistical output.

The program that we have often recommended to students is "G*Power," which can be downloaded at www.psycho.uni-duesseldorf.de/aap/projects/gpower/.

Other Resources for Calculating Power

If you have an experimental design that does not fit with the options for calculating power with G*Power, you can use the following website to find more free resources on the Internet for calculating statistics, including power: http://statpages.org/.

Notice that the flowchart (see page 214) does not yet tell you what to do if you do not meet the assumptions of your statistical tests. In fact, there are some options when you have three samples, but you do not meet the assumptions for parametric tests. The one-way ANOVA can be replaced with a nonparametric test called the Kruskal-Wallis test. Another option is to perform a log transformation on your data to determine if the mathematical transformation will result in a normal distribution.

SUMMARY

In this chapter, you have covered statistical tests for designs in which you have three or more samples in an independent, or between-groups, design, and you lack information about the underlying population. We have shown you the steps involved in conducting your overall analyses (F test) as well as the steps involved in determining where your means are significantly different (providing that your first overall F test is significant). In the next chapter, we will extend the concept of an independent one-way ANOVA design to more complex situations where we have a within-groups design or more than one factor (independent variable).

Excel Step-by-Step: Step-by-Step Instructions for Using Microsoft Excel 2003 or 2007 to Run ANOVA

1. Your first step will be to open Microsoft Excel and type the raw data into a spreadsheet (data listed below). It is helpful to type the column headers so that your output will be labeled later. Note that the participants only received one of the three different doses.

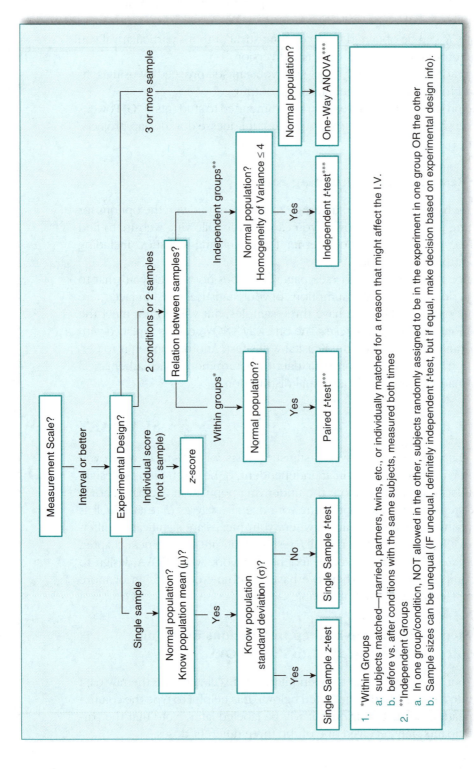

Overview of Single-Sample, Two-Sample, and Three or More Sample Tests: Expanded Flowchart

The flowchart contains the following elements:

Measurement Scale? → Interval or better → Experimental Design?

From Experimental Design?:
- Single sample → Normal population? Know population mean (μ)? → Yes → Know population standard deviation (σ)? → Yes → Single Sample z-test; No → Single Sample t-test
- Individual score (not a sample) → z-score
- 2 conditions or 2 samples → Relation between samples?
  - Within groups* → Normal population? → Yes → Paired t-test***
  - Independent groups** → Normal population? Homogeneity of Variance ≤4 → Yes → Independent t-test***
- 3 or more sample → Normal population? → One-Way ANOVA***

1. *Within Groups
   a. subjects matched—married, partners, twins, etc., or individually matched for a reason that might affect the I.V.
   b. before vs. after conditions with the same subjects, measured both times
2. **Independent Groups
   a. In one group/condition, NOT allowed in the other, subjects randomly assigned to be in the experiment in one group OR the other
   b. Sample sizes can be unequal (IF unequal, definitely independent t-test, but if equal, make decision based on experimental design info).

0 mL	100 mL	300 mL
15	18	26
17	19	30
19	22	31
23	27	36
20	21	29

2. Once your data are entered, you will need to calculate an ANOVA.

For Excel 2003: To find the *t* test you will need, you can go to the built-in data analysis function. You'll find the option under the "Tools" menu, and at the bottom of the list that pops up, you should see "Data Analysis."

If you do not see the "Data Analysis" option under the Tools menu, select "Add Ins" under the Tool menu. Check the box next to "Analysis TookPak" and click OK. Follow any further instructions that the computer gives you.

For Excel 2007: To get to the data analysis option, click on the "DATA" tab, and the "Data Analysis" tool will be in the "Analysis" section to the far right of the screen. Once you find the Data Analysis tool, the rest of the instructions are the same.

3. After you click on "Data Analysis," a list of possible statistical tests will pop up. Page down the list until you find the appropriate test. We want to do a one-way ANOVA, which Excel calls "ANOVA: Single Factor." After you have selected the appropriate test, the program will take you through the steps to complete the test. For example, input your range (highlight range of data with your mouse) and also indicate your alpha level (usually 0.05).

Here is example output from an ANOVA:

ANOVA: Single Factor				
Summary				
Groups	**Count**	**Sum**	**Average**	**Variance**
0 ml	5	94	18.8	9.2
100 ml	5	107	21.4	12.3
300 ml	5	152	30.4	13.3

(Continued)

(Continued)

ANOVA						
Source of Variation	SS	df	MS	F	P-value	F crit
Between Groups	370.5333333	2	185.2666667	15.97126437	0.000414741	3.885290312
Within Groups	139.2	12	11.6			
Total	509.7333333	14				

Note: p value is less than 0.05 (or F obt > F crit); therefore, reject the null hypothesis.

SPSS Step-by-Step: Step-by-Step Instructions for Using SPSS to Run a One-Way ANOVA

1. Your first step will be to open SPSS and select the option that allows you to type in new data.

2. This will open a page called "Variable View." To confirm that, look at the tab at the bottom left of the page. There should be two tabs, and one will say "Variable View" (the one you are in now), and the other will say "Data View."

3. Now you need to establish your variables for SPSS. Make a variable that codes for your independent groups by typing in the word *group* in the first box of row 1. Now name your dependent variable and call it *dv* and type that into box 1 in row 2. By default, SPSS will consider each of these variables to be numeric, and for these purposes, all of the default codes will work perfectly. However, keep in mind that this is where you can change some of your options to allow for alphabetical data, define coding in your variables, and so on. For example, you could define your group coding to be 1 for the control group and 2 for one level of your experimental treatment group and 3 for another level of your experimental treatment group (for perhaps another dose level).

4. Click on the "Data View" tab now. You should see that the variable names you entered in Variable View have now appeared at the top of this spreadsheet. Now you can enter your raw data:

Group	dv
1	15
1	17
1	19
1	23
1	20
2	18
2	19
2	22
2	27
2	21
3	26
3	30
3	31
3	36
3	29

5. From the SPSS menu, you should now select "Analyze," then "Compare means," and finally "One-way ANOVA." This will open up a new pop-up window with your variables listed on the left-hand side. Select your dependent variable and use the arrow in the middle of the pop-up to move it to the right-hand side of the pop-up box as your "dependent variable." Now move your "group" variable the same way over to the "factor" box.

6. For the short version of output, you can just hit "OK" to run the test. You'll get the output for an ANOVA summary table that looks familiar!

7. For the long version of the output, select options *before* hitting OK. Under options, you should select descriptive statistics and homogeneity of variance (HOV). Then select "Continue" and then "OK" to run the test.

8. With this longer option, you will get the descriptive statistics, Levene's test for homogeneity of variance (to see if your independent groups meet the homogeneity of variance assumption), and an F summary table.

CHAPTER 10 HOMEWORK

Provide a short answer for the following questions.

1. How does variability of the samples, in all of its forms, influence the F-obtained value?

2. The F distribution has no negative values. True or false? How do you account for this?

3. Why do we have to use Tukey's test instead of multiple independent t tests?

4. If $F < 1$, we know that there was no significant effect of our independent variable. How do we know this?

5. How is the F distribution similar to the t distribution?

6. What are the assumptions for one-way ANOVA?

7. How do you check homogeneity of variance when conducting an ANOVA test?

8. A clinical researcher tests the effects of three drugs on motor coordination and obtains motor coordination scores for each participant. There are three groups with four participants in each condition. Assume that the population of scores is normally distributed and all variances are homogeneous. Complete the following ANOVA source table. Using an $\alpha = 0.05$, what do you conclude? Reject or fail to reject the H_0 and perform Tukey's test, if appropriate.

Drug A: $\bar{X} = 9$

Drug B: $\bar{X} = 9.5$

Drug C: $\bar{X} = 15.25$

Source	SumsSquares	df	s^2	F_{obt}	F_{crit}
Between 148.17	____	____	____	____	
Within	____	____	7.1944		
Total	____	____			

9. A research team specializing in borderline personality disorder tests three new drugs to treat the condition. Eighty-four volunteers with borderline personality disorder are divided equally and randomly into four groups. Groups A to C each receive different experimental drugs to treat the disorder along with behavioral therapy, and Group D receives a placebo and behavioral therapy. Levels of a neurotransmitter known to be associated with the disorder are

assessed after 6 months of therapy, and the groups are compared. Assume that the population of scores is normally distributed and all variances are homogeneous. Test the hypothesis that the drug affects neurotransmitter levels. Complete the following analysis of variance table, reject or fail to reject the H_0, and conduct Tukey's test if appropriate:

Group A: $\bar{X} = 155.25$

Group B: $\bar{X} = 123.33$

Group C: $\bar{X} = 59.25$

Group D: $\bar{X} = 165.99$

Source	SumsSquares	df	s^2	F_{obt}	F_{crit}
Between	3.2661	_____	_____	_____	_____
Within	_____	_____	_____		
Total	40.5061	_____			

Choose the most appropriate and powerful test:

10. Several studies indicate that handedness (left-handed/right-handed) is related to differences in brain function. Because different parts of the brain are specialized for specific behaviors, this means that left-handed and right-handed people should show different skills or talents. To test this hypothesis, a psychologist tested pitch discrimination for three randomly selected groups of participants: 34 left-handed, 31 right-handed, and 30 ambidextrous. Pitch discrimination is assessed based on the participant's ability to determine if the chord is a third (notes separated by four half steps) or a fifth (notes separated by seven half steps). Variances are 1.8, 2.4, and 3.2, respectively.

11. Research on human memory indicates that, on average, people can remember five pieces of new information at a time. A random sample of 16 participants from a normally distributed population trained in mnemonics were read a list of 20 words to remember, asked to count backwards from 100 for a minute, and then asked to recall as many of the words as possible. Their mean recall was 8 words with a standard deviation of 2 words. Using this information, determine whether the participants in this sample were representative of the general population in which the average recall is known to be 5 words.

12. Over a 20-year period, first-year students at a state university had a mean grade point average (GPA) of 2.27 with a standard deviation of 0.42. Recently, the university has tried to increase admissions standards. A sample of 30 first-year students accepted into the university this year had a mean GPA of 2.63. Have the new admissions standards resulted in a significant change in GPA of first-year students?

13. A researcher is interested to know whether there is any difference in mathematical aptitude for 6-year-old boys and 6-year-old girls. Thirty-five children of both sexes were randomly selected to participate in the study. Both groups were given a math skills test in which the number of questions answered correctly was evaluated. Girls averaged a mean of 23 problems right with a standard deviation of 4. Boys averaged a mean of 27 problems right with a standard deviation of 6. Can the researcher conclude that 6-year-old boys and girls differ in their mathematical aptitude?

Answer Questions 14 to 17 using the following data.

A behavioral scientist wishes to compare the effectiveness of five different weight loss programs. Six participants are randomly assigned to each of the programs, and weight loss per month is recorded.

Programs	1	2	3	4	5
Means	6.33	9.50	8.67	14.33	3.00

14. State the hypotheses.

15. Complete the following ANOVA summary table.

Source	SS	df	s^2	F_{obt}	F_{crit}
Between			104.87		
Within	169.49				
Total					

16. Decide whether to reject the null hypothesis.

17. Perform post hoc testing if appropriate and state your conclusion.

Answer Questions 18 to 22 using the following story problem and data.

Dr. Snooze is interested in the effect of sleep deprivation on problem-solving ability. She measures the time in minutes it takes to solve a statistical story problem, comparing three groups of people. The first group has been awake for 20 hours, the second group has been awake for 40 hours, and the third group has been awake for 60 hours. The time it takes to solve these types of problems is normally distributed, and the variances are homogeneous. The first group of four people takes an average of 2 minutes to solve the problem, the second group of four people takes an average of 3.5 minutes to solve the problem, and the third group of four people takes an average of 5 minutes to solve the problem.

18. State the null hypothesis and alternative hypothesis.

19. Complete the following ANOVA summary table.

Source	SS	df	s^2	F_{obt}	F_{crit}
Between					
Within	19		2.11		
Total	35				

20. Decide whether to reject the null hypothesis.

21. Perform post hoc testing if appropriate.

22. State the conclusion.

For Questions 23 to 24, use the following story problem and data.

A pool of participants was randomly divided into three treatment groups. The groups were administered daily doses of vitamin C over a 12-month period. The data in the table represent the number of cold and flu viruses reported by the participants as a function of their vitamin C dosage.

Raw Scores

0 mg	500 mg	2000 mg
6	3	1
5	3	0
3	4	2
2	2	1

Means

0 mg	500 mg	2000 mg
4	3	1

23. Complete the following ANOVA summary table using the .05 level of significance.

Source	SS	df	S^2	F_{obt}	F_{crit}
Between	18.72	2			
Within	14	9	1.56	X	X
Total	32.72	11	X	X	X

24. Using the story problem on vitamin C and the raw data in the table above, calculate an ANOVA using the raw data. Show your work for full credit.

The following homework questions should be answered with the online data set provided for this chapter via the textbook's website.

25. Produce a table of descriptive statistics using Microsoft Excel or SPSS.

26. Interpret the descriptive statistics produced by Excel. Do you meet the assumption of normality?

27. Do you meet the assumption of homogeneity of variance?

28. Analyze the data set using a one-way ANOVA, and indicate if you have a statistically significant result. Explain how you evaluated the output.

CHAPTER 11
Complex ANOVA Designs

"Complex ANOVA designs" may sound intimidating, but it's really just an extension of the last chapter. Some of the modifications involve adding another independent variable (factor) to the design, while others are adjustments that you make when you analyze two or more "paired" or repeated or within-groups (same subjects at different times, for example). These kinds of analyses may be relevant if you have an experimental design where you assess the developmental stages of children at ages 2 years, 4 years, and 8 years. Alternatively, you might be interested in complex interactions between age (in years), as well as gender and socio-economic background. In Renee Ha's work with animals, she is frequently including factors such as biological sex, breeding season, and age class into understanding behavior in the field.

In our discussion of analysis of variance (ANOVA) in the last chapter, we suggested that each score is the result of adding to the grand mean (μ, the overall mean of all of the scores regardless of group membership) the combined influence of any treatment effect attributable to our independent variable (IV) and the unexplained error (E). This was represented by a linear or additive equation:

$$\text{Score} = \mu + \text{IV} + \text{E}.$$

In this chapter, we will modify that conceptual equation to account for additional independent variables (or factors) and for situations in which you use a within-groups experimental design.

WHAT IS A FACTORIAL DESIGN?

The simplest factorial design is a 2×2 design. In the logic of research methods, this involves two independent variables, each with two levels. For example, you could be studying the influences of sex (male vs. female) on two levels of a treatment drug (low vs. high; see Table 11.1). Note that factorial designs can involve more levels of each independent variable and thus become a 2×3 design, a 3×3 design, and so on. We provide further examples later in the chapter of these more complex designs (see the discussion on multifactorial ANOVA in particular). Advantages of the factorial design include more information for the same amount of effort (e.g., another variable/ factor evaluated), the ability to understand how a second variable/factor influences the variability in your first factor (potentially explaining more of the error variance), and the opportunity to study interactions between factors.

Table 11.1		Factor B (Sex)	
		Male	Female
Factor A (dose)	Low	$n = 25$	$n = 25$
	High	$n = 25$	$n = 25$

Simple Factorial Design (2×2)

HOW TO ANALYZE TWO-FACTOR DESIGNS AND INTERACTIONS

What do you do if you have more than one independent variable in your experiment that you need to analyze? Your first thought might be that you could perform two one-way ANOVAs or one ANOVA for each independent variable. The problem with this approach is the same one that we faced when we discussed doing multiple independent t tests when you have three or more groups: the issue of Type I error. The more tests you perform, the greater your probability of making a Type I error (rejecting the null hypothesis when it is, in fact, true). The solution to the problem is to simultaneously analyze your independent variables in one overall analysis (F test). To do that, we need a single test for two independent variables, and under the correct conditions, the test that fits those requirements is a two-way ANOVA. One of the additional advantages of using this test is that you can evaluate whether there is an interaction between your two independent variables. This concept of interactive effects is a new one.

What Is an Interaction?

When the effect of one of your independent variables (or factors) is not the same at all levels of your second independent variable, then you have a statistical "interaction" effect. In other words, the effect is not additive or linear. For example, a higher value of Factor A may lead to a higher value of Factor B, but a lower value of Factor A may lead to a higher value of Factor B, and moderate values of A may also lead to lower values of B. The influence of Factors A and B on a dependent variable are not linear and additive but still have a significant effect. This can make explaining the relationships between your factors more complicated. However, real-world examples may involve interactions. For example, treatment of depression may have the largest benefit from both medication and counseling, and the improvement may be over and above the effects of medication or counseling alone. When alleviation of depression symptoms is greatest for both treatments, over and above the improvement seen by each treatment, then you have an interaction. In other words, medication may create a situation whereby the client is better able to be helped by counseling. See the example on hesitation time and Figure 11.1 for another explanation of statistical interactions.

Applying the Two-Way ANOVA Concept to Our Hesitation Time Example

If we continue to use the hesitation time example with which you are familiar from the last chapter, then making that design a two-factor design could be done as easily as adding information on sex for each participant. Thus, we would be asking, "What are the effects of sex and of ethnic background on the hesitation time?" One factor would be sex, and all of the data would be averaged based on sex alone (male or female), while the other factor would be ethnic group, and the influence of ethnic group (A or B or C) would be calculated while ignoring sex. As a result of this analysis, you can get an idea of the relative influence of one factor relative to another factor. These influences are technically termed **main effects,** so that you have a "main effect" for ethnicity and a "main effect" for sex. Sometimes the influences of the two independent variables are completely independent of one another, but sometimes they are not. When the effect of one of your independent variables (or factors) is not the same at all levels of your second independent variable, then you have a statistical **interaction** effect. Examples of main and interaction effects are illustrated in Figure 11.1. A significant interaction occurs when there is an effect of the combined influence of your independent variables that is over and above their individual independent effects.

> **Main effect:** Influence of your independent variable on your dependent variable.

> **Interaction:** The combined effect of both (or all) independent variables over and above the separate effects of each variable alone. This is often reflected in the fact that one of your independent variables affects the dependent variable differently at different levels of your second independent variable.

In Figure 11.1a,b, the effects of sex are the same for each ethnic group (in 11.1a, males have longer hesitation times across all ethnic groups, and in 11.1b, there is no difference between sexes across all ethnic groups). However, in Figure 11.1c, there is no sex difference in Group A but a significant difference between sexes in Group C (sex differences in Group B might, or might not, be significant). This would represent a significant interaction effect. The significant interaction effect is similar in Figure 11.1d, except that there is clearly no sex effect in Group A or Group B. The significant interaction effect in Figure 11.1e is most dramatic since the effect of sex is in the *opposite direction* in Group A and Group C, and there is no sex difference in Group B.

Linear or Additive Equation

There are multiple potential influences (e.g., ethnic group or sex) on the outcome for each score (e.g., an individual participant's hesitation time), and those influences can be defined with a linear equation. While we will not directly use this formula in this chapter, we will discuss this a bit more in Chapter 13. For now, it is important to focus on whether our treatment effects are enough to demonstrate a statistically significant

Figure 11.1

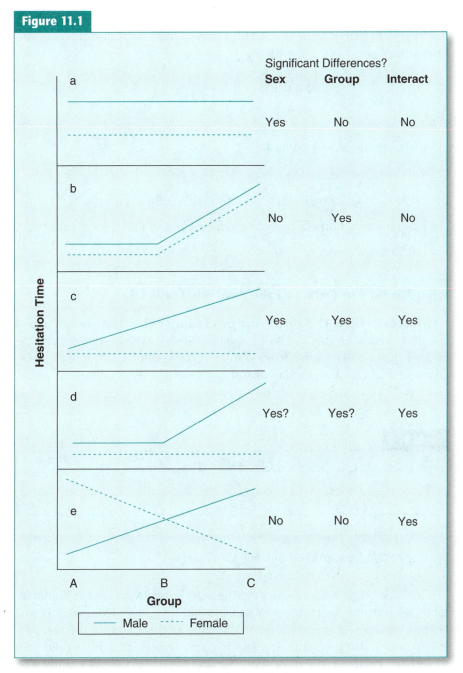

Examples of Main Effects and Interactions in Graphical Form

difference between our groups. Conceptually, an individual's score (in this case, one participant's hesitation time) could be represented in the following way:

$$\text{Score} = \mu + \text{IV}_1 + \text{IV}_2 + (\text{IV}_1)(\text{IV}_2) + \text{E},$$

where

μ = the grand mean starting point,

IV_1 = effect of the first independent variable (sex),

IV_2 = effect of the second independent variable (ethnic group),

$(\text{IV}_1)(\text{IV}_2)$ = interaction effect of the independent variables (Sex × Group),

E = unexplained error.

What Would the Data and the Analysis Look Like?

Notice in Table 11.2 that using two factors rather than one typically results in more groups, or more categories, into which subjects can be divided. If we were to break up our 15 participants by sex as well as by group, it might look something like this. Obviously, this sample size would be considered extremely small for this analysis.

Table 11.2

		Ethnic Group		
		A	B	C
Sex	Male	$n = 2$	$n = 3$	$n = 3$
	Female	$n = 3$	$n = 2$	$n = 2$

Two-Way ANOVA Sample Sizes (Hesitation Time Example)

In Table 11.3, we have provided a sample two-way ANOVA summary table. Both main effects (the row variable and the column variable) as well as the interaction are represented in the table. In this example, the row variable is sex and the column variable is ethnic group, but this would vary depending on the two independent variables you used in your study. The interaction is the combination of sex and group over and above any independent effect they might have on hesitation time.

Table 11.3

Source	SS	df	S² (MS)	F Obtained
Row (sex)				
Column (group)				
Row × Column (Sex × Group)				
Within (within a cell)				
Total				

Two-Way ANOVA Summary Table (Hesitation Time Example)

This two-factor ANOVA would produce three F-obtained values. One of them would be for a main effect of sex, one would be the main effect of ethnic group, and the last value would indicate whether there was a significant interaction of sex and ethnic group over and above their main effects (i.e., after accounting for the variance attributed to the main effects). These F values would be calculated by dividing the variance attributed to a main effect (i.e., row variance or s_R^2) by the within-cell variance (s_w^2) and then doing that again for the other main effect (i.e., column variance or s_C^2) and for the interaction term (s_{RC}^2). Mathematically, that would look like this:

$$F \text{ obtained for row factor (sex)} = \frac{s_R^2}{s_w^2}.$$

$$F \text{ obtained for column factor (group)} = \frac{s_C^2}{s_w^2}.$$

$$F \text{ obtained for Row} \times \text{Column (interaction)} = \frac{s_{RC}^2}{s_w^2}.$$

s_R^2 = Between-row variance (reflects natural variation in the population and any variability due to your row variable).

s_w^2 = Within-cell variance (reflects natural variation in the population, and thus it is an estimate of σ^2). Within-cell variance could also be called "error" (E) because it is variability that we cannot control but occurs naturally.

s_C^2 = Between-column variance (reflects natural variation in the population and any variability due to your column variable).

s_{RC}^2 = Row × Column variance (reflects natural variation in the population and any variability due to the interaction between your row and column variables).

Complete Example

Craik and Lockhart (1972) proposed a model of memory that suggested that the degree to which verbal material is remembered by the subject is a function of the degree to which it was processed when it was initially presented. To test this model, a researcher randomly assigned 50 participants between the ages of 55 and 65 and a second group of 50 participants between the ages of 21 and 25 to one of five groups: four incidental learning groups (without the expectation that the material will need to be recalled later) and one intentional learning group. The incidental groups were asked to process the words in various ways but were not told that they would later have to re-create the list of words. The intentional group was told to read through the list and to memorize the words for later recall. After participants had gone through the list of 27 items three times, they were given a sheet of paper and asked to write down all of the words they could remember. The success of each group was compared using a two-way ANOVA. Age and memorization technique were the two factors (or independent variables), and the number of words from a list that were later recalled was the dependent variable. Figure 11.2 shows these data graphically. Complete the two-way ANOVA summary table (Table 11.4) based on the formulas for the three F-obtained values you were given above, and evaluate whether or not age and memorization techniques as well the interaction between them have a significant effect on the words correctly recalled. Assume that the data are normally distributed and the variances are homogeneous. Use $\alpha = 0.05$.

Null hypothesis (main effect for age): There is no difference in the number of words recalled based on age effects or the differences are due to chance.

Alternative hypothesis (main effect for age): There is a difference in the number of words recalled based on age effects.

Null hypothesis (main effect for technique): There is no difference in the number of words recalled based on technique or the differences are due to chance.

Alternative hypothesis (main effect for technique): There is a difference in the number of words recalled based on the technique employed.

Null hypothesis (interaction Age × Technique): There is no difference in the number of words recalled based on the interaction between age and technique or the differences are due to chance.

Alternative hypothesis (interaction Age × Technique): There is a difference in the number of words recalled based on the interaction between age and technique.

Figure 11.2

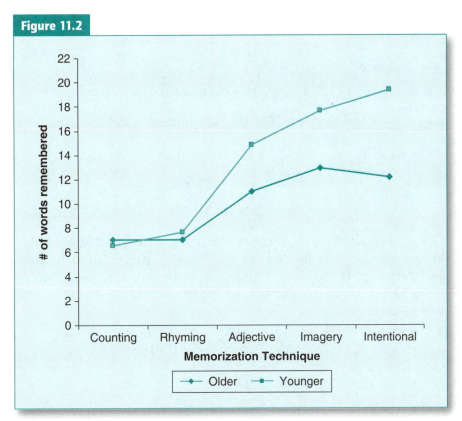

Craik and Lockhart (1972) Memorization Experiment. The top line represents the number of words recalled by younger individuals, and the bottom line represents the number of words recalled by older individuals. Note that the results are similar between age groups for the counting and rhyming techniques.

Table 11.4

Source	SS	df	S² (MS)	F Obtained
Row (age)	249.64	1		
Column (technique)	1476.66	4		
Row × Column (Age × Technique)	195.26	4		
Within (within a cell)	628.80	90		
Total				

Two-Way ANOVA Summary Table (Memory Example)

★ **Step 1:** Apply the s^2 and F-obtained formulas to complete the ANOVA summary table.

Calculate SS_T:

$$SS_T = SS_R + SS_C + SS_{RC} + SS_W.$$

$$SS_T = 249.64 + 1476.66 + 195.26 + 628.80 = 2550.36.$$

Calculate df_T:

$$df_T = df_R + df_C + df_{RC} + df_W.$$

$$df_T = 1 + 4 + 4 + 90 = 99.$$

Calculate variances: Calculate variance the same way that you did in the one-way ANOVA, by dividing the sums of squares (SS) by their appropriate df. If you have the table laid out properly, then you can divide the first column of numbers by their respective df (second column of numbers).

$$s_R^2 = \frac{SS_R}{df_R} = \frac{249.64}{1} = 249.64.$$

$$s_C^2 = \frac{SS_C}{df_C} = \frac{1476.66}{4} = 369.165.$$

$$s_{RC}^2 = \frac{SS_{RC}}{df_{RC}} = \frac{195.26}{4} = 48.815.$$

$$s_w^2 = \frac{SS_W}{df_W} = \frac{628.80}{90} = 6.9833337.$$

Calculate F-obtained values:

$$F\text{-obtained for row factor (age)} = \frac{s_R^2}{s_w^2} = \frac{249.64}{6.9866667} = 35.73.$$

$$F\text{-obtained for column factor (technique)} = \frac{s_C^2}{s_w^2} = \frac{369.165}{6.9866667} = 52.84.$$

$$F\text{-obtained for Row} \times \text{Column (interaction)} = \frac{s_{RC}^2}{s_w^2} = \frac{48.815}{6.9866667} = 6.99.$$

Step 2: Evaluate the probability of obtaining this score due to chance.

Evaluate the *F*-obtained values based on alpha (α) = 0.05 and a two-tailed hypothesis. To evaluate your *F*-obtained value, you must use the *F* distribution (Table C in the Appendix). To determine your *F*-critical value, you need to know your alpha level (0.05) and your degrees of freedom (df_B and df_W). The degrees of freedom from the numerator of the *F*-obtained equation are the df_B, while the degrees of freedom from the denominator are df_W. Thus, the *df* for this problem are 1 and 90, 4 and 90, and 4 and 90, respectively, for the three *F* values you have calculated. Compare the appropriate *F*-critical value with the appropriate *F*-obtained value, and reject the null hypothesis when your *F*-obtained value is equal to or greater than your *F*-critical value. You should use 3.95, 2.47, and 2.47 as your *F*-critical values.

F obtained ≥ *F* critical in all three cases, so we *reject* the null hypotheses for both main effects (of age and technique), and we reject the null hypothesis for the interaction term.

How should we interpret these data in light of the effect of age and memorization technique on the ability to recall words from a list correctly?

These results suggest that age and technique have significant effects on the ability to recall a list of words and that there is some interaction between age and technique that has effects above and beyond the influences of each separately. Based on the figure alone, it appears that younger adults recall more words than older adults and that "adjective," "imagery," and "intentional techniques" are most successful. The "intentional" strategy is particularly successful in younger adults (interaction term). However, these conclusions are based purely on the appearance of the figure and knowledge that the overall *F* tests are significant. Proper multiple comparisons would need to be conducted to determine which cells are significantly different from which other cells. As we discussed in Chapter 10, determining the statistical significance of these pairwise comparisons can only be done using "multiple comparisons" once you know that your overall test (your ANOVA) has reached statistical significance. Take a look at the Excel and SPSS output for these data that we have manually analyzed.

Results if you use Microsoft Excel to calculate the two-way ANOVA:

Source of Variation	SS	Df	MS	F	P-value	F crit
Sample	249.64	1	**249.64**	**35.73091603**	4.4838E-08	**3.946865945**
Columns	1476.66	4	**369.165**	**52.83850191**	7.85493E-23	**2.472930305**
Interaction	195.26	4	**48.815**	**6.986879771**	6.04944E-05	**2.472930305**
Within	628.8	90	**6.9866667**			
Total	**2550.36**	**99**				

ANOVA: Two-Factor

Note that Excel calls this test an "ANOVA: Two Factor." This wording is slightly different from what we have been using, but it is describing the same analysis. We have bolded the numbers that are comparable to the numbers we manually calculated or looked up in the table. Recall that another name for variance is mean square (MS), and Excel uses an MS label in the table. Note that the F-obtained values (Excel calls it "F") are identical to ours, except they have carried more digits. The "F Crit" values are also the same, but again they have carried more digits. The degrees of freedom are the same. In addition, they have provided you with the exact probability ("P-value") of obtaining each F-obtained value ("F") of which is shown in scientific notation. The notation "4.4838E-08" tells us that we should move the decimal place left 8 places, while "7.85493E-23" means that the decimal will have to be moved to the left 23 places. Thus, in all three cases, all of the probabilities are very small and definitely < 0.05, so we clearly *reject all* of our null hypotheses.

Results if you use SPSS to calculate the ANOVA (or F test):

Source	Type III Sum of Squares	df	Mean Square	F	Sig.
Model	15331.200	10	1533.120	219.435	.000
AGE	249.640	1	**249.640**	**35.731**	.000
TECH	1476.660	4	**369.165**	**52.839**	.000
AGE * TECH	195.260	4	**48.815**	**6.987**	.000
Error	628.800	90	**6.987**		
Total	15960.000	100			

Tests of Between-Subjects Effects

Dependent Variable: SCORE

The first F value in the table is an overall F test that combines main effects and the interaction in one analysis. This is like the one-way ANOVA you did in Chapter 10. The remaining F values are just like the Excel printout except that there are no F-critical values provided, and the p value is now indicated as "Sig." If the "significance" value is less than your alpha level (which is usually 0.05), then you can reject the null hypothesis.

How would these results be reported in a scientific journal article?

If the journal required American Psychological Association (APA) format, the results would be reported in a format something like this:

There was a significant difference in the number of words recalled based on age effects, $F(1, 90) = 35.73$, $p < .01$. In general, younger individuals recalled more words than older individuals.

There was a significant difference in the number of words recalled based on the technique employed, $F(4, 90) = 52.84$, $p < .01$. Counting and rhyming were less effective than all other techniques.

There was a significant interaction between age and technique that affected the number of words recalled, $F(4, 90) = 6.99$, $p < .01$. Younger individuals recalled more words when using adjective, imagery, and particularly intentional techniques compared to older individuals.

These formal sentences include the dependent variable (words recalled), the independent variables (age, technique, and the interaction between age and technique), as well as a statement about statistical significance, the symbol of the test (F), the degrees of freedom between groups (1 or 4), the degrees of freedom within groups (90), the statistical values (35.73, 52.84, and 6.99), and the probability of obtaining these results simply due to chance (all $< .01$).

Another Complete Example

A researcher examined the outcomes of social interactions between two individuals and hypothesized that the rate of very subtle eye glances toward the other interactor would predict the "winner" of the interaction. But first she wanted to answer a basic question: Did the rates of eye glances differ based on the gender of the interactors and the type of discussion involved ("happy," "controversial," or "neutral" topics)? The researcher placed independent male-male, male-female, and female-female pairs of individuals into one of the three discussion types and coded the number of eye glances recorded in the session as a whole. The differences among the groups could be compared using a two-way ANOVA. Pair gender and discussion type were the two factors (or independent variables), and the number of eye glances was the dependent variable. Figure 11.3 shows these data graphically. Complete the two-way ANOVA summary table (Table 11.5) based on the formulas for the three F-obtained values you were given above, and evaluate whether pair gender and discussion type, as well the interaction between them, have a significant effect on the number of eye glances. Assume that the data are normally distributed and the variances are homogeneous. Use $\alpha = 0.05$.

Null hypothesis (main effect for pair gender): There is no difference in the number of eye glances based on pair gender (male-male vs. male-female vs. female-female), or the differences are due to chance.

Alternative hypothesis (main effect for pair gender): There is a difference in the number of eye glances due to the effects of pair gender.

Null hypothesis (main effect for discussion type): There is no difference in the number of eye glances based on discussion type, or the differences are due to chance.

Alternative hypothesis (main effect for discussion type): There is a difference in the number of eye glances due to the effects of discussion type.

Null hypothesis (interaction: pair gender by discussion type): There is no difference in the number of eye glances based on the interaction between pair gender and discussion type over and above the differences that are due to the main effects of pair gender and discussion type, or the differences are due to chance. There is an alternate form of this hypothesis: There is no difference in number of eye glances among the individual cell means, or the differences are due to chance.

Alternative hypothesis (interaction: pair gender by discussion type): There is a difference in the number of eye glances due to the interaction between pair gender and discussion type. Alternatively, there is a difference among one or more of the cell means.

Figure 11.3

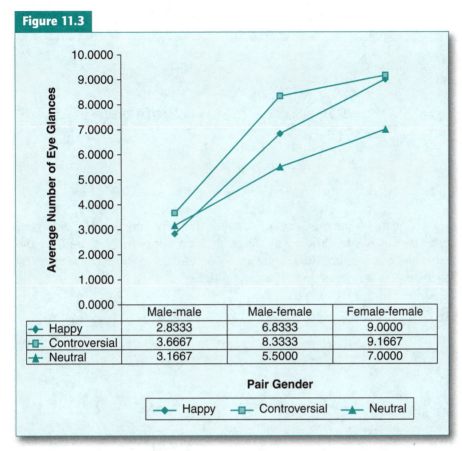

	Male-male	Male-female	Female-female
Happy	2.8333	6.8333	9.0000
Controversial	3.6667	8.3333	9.1667
Neutral	3.1667	5.5000	7.0000

Pair Gender

Happy — Controversial — Neutral

Pair gender (*X*-axis) and discussion type (lines) were the two factors (or independent variables), and the number of eye glances was the dependent variable (shown here on the *Y*-axis). Note that the average number of eye glances appears to be lowest for male-male pairs, but let's check that statistically.

Table 11.5

Source	SS	df	S^2 (MS)	F Obtained
Row (discussion type)	30.33333	2		
Column (pair gender)	254.3333	2		
Row × Column (Discussion Type × Pair Gender)	13.33333	4		
Within (within a cell)	349.5	45		
Total				

Two-Way ANOVA Summary Table (Eye Glance Example)

★ **Step 1:** Apply the s^2 and F-obtained formulas to complete the ANOVA summary table.

Calculate SS_T:

$$SS_T = SS_R + SS_C + SS_{RC} + SS_W.$$

$$SS_T = 249.64 + 1476.66 + 195.26 + 628.80 = 2550.36.$$

Calculate df_T:

$$df_T = df_R + df_C + df_{RC} + df_W.$$

$$df_T = 1 + 4 + 4 + 90 = 99.$$

Calculate variances: Calculate variance the same way that you did in the one-way ANOVA, by dividing the SS by their appropriate df. If you have the table laid out properly, then you can divide the first column of numbers by their respective df (second column of numbers).

$$s_R^2 = \frac{SS_R}{df_R} = \frac{30.3333}{2} = 15.1667.$$

$$s_C^2 = \frac{SS_C}{df_C} = \frac{254.3333}{2} = 127.1667.$$

$$s_{RC}^2 = \frac{SS_{RC}}{df_{RC}} = \frac{13.33333}{4} = 3.3333.$$

$$s_w^2 = \frac{SS_W}{df_W} = \frac{349.5000}{45} = 7.7667.$$

Calculate F-obtained values:

F-obtained for row factor (discussion type) =

$$\frac{s_R^2}{s_w^2} = \frac{15.1667}{7.7667} = 1.9528.$$

F-obtained for column factor (pair gender) =

$$\frac{s_C^2}{s_w^2} = \frac{127.1667}{7.667} = 16.3734.$$

F-obtained for Row × Column (interaction) $= \dfrac{s_{RC}^2}{s_w^2} = \dfrac{3.3333}{7.7667} = 0.4292.$

Step 2: Evaluate the probability of obtaining this score due to chance.

Evaluate the F-obtained values based on alpha (α) = 0.05 and a two-tailed hypothesis. To evaluate your F-obtained value, you must use the F distribution (Table C in the Appendix). To determine your F-critical value, you need to know your alpha level (0.05) and your degrees of freedom (df_B and df_W). The degrees of freedom from the numerator of the F-obtained equation are the df_B, while the degrees of freedom from the denominator are df_W. Thus, the df for this problem are 2 and 45, 2 and 45, and 4 and 45, respectively, for the three F values you have calculated. Compare the appropriate F-critical value with the appropriate F-obtained value, and reject the null hypothesis when your F-obtained value is equal to or greater than your F-critical value. You should use 3.21, 3.21, and 2.58 as your F-critical values.

F obtained $\geq F$ critical only in the case of pair gender, so we *reject* the null hypothesis for the main effect of pair gender and *fail to reject* the null hypotheses for the main effect of discussion type and interaction.

How should we interpret these data in light of the effect of pair gender and discussion type on the number of eye glances?

These results suggest that pair gender has a significant effect on the number of eye glances, as can be seen in Figure 11.3: Male-male pairs have relatively fewer eye glances, regardless of the discussion type, while male-female and female-female pairs exhibit increasing numbers of eye glances, again regardless of the discussion type. When discussion type alone is considered, there is no significant difference in number of eye glances, and there is no interactive effect of pair gender and discussion type over and above the main effects already described. However, the direction of the pair gender effect is based entirely on the appearance of the figure and the knowledge that the overall F test for pair gender is significant. Proper multiple comparisons would need to be done to determine which cells are significantly different from which other cells. As we discussed in Chapter 10, determining the statistical significance of these pairwise comparisons can only be done using multiple comparisons once you know that your overall test (your ANOVA) has reached statistical significance. Take a look at the Excel and SPSS output for these data that we have manually analyzed.

Results if you use Microsoft Excel to calculate the two-way ANOVA:

Source of Variation	SS	df	MS	F	P-value	F crit
Sample	30.33333	2	**15.16667**	**1.95279**	0.153715	**3.20432**
Columns	254.3333	2	**127.1667**	**16.37339**	4.54E-06	**3.20432**
Interaction	13.33333	4	**3.333333**	**0.429185**	0.786769	**2.578737**
Within	349.5	45	**7.766667**			
Total	**647.5**	**53**				

ANOVA

We have bolded the numbers that are comparable to the numbers we just manually calculated or looked up in the table. Recall that another name for variance is mean square (MS), and Excel uses an MS label in the table. Note that the *F*-obtained values (Excel calls it "F") are identical to ours, except they have carried more digits. The "F Crit" values are also the same, but again they have carried more digits. The degrees of freedom are the same. In addition, they have provided you with the exact probability ("P-value") of obtaining each *F*-obtained value ("F") of which is shown in scientific notation. The notation "4.54E-06" tells us that we should move the decimal place left six places (0.00000454). Thus, only in the case of the pair gender effect is the probability small (i.e., < 0.05), and so we only *reject* that null hypothesis.

Results if you use SPSS to calculate the ANOVA (or *F* test):

Source	Type III Sum of Squares	df	Mean Square	F	Sig.
Model	298.000	8	37.250	4.796	.000
DISCTYP	30.333	2	**15.167**	**1.953**	.154
PAIRGEN	254.333	2	**127.167**	**16.373**	.000
DISCTYP * PAIRGEN	13.333	4	**3.333**	**.429**	.787
Error	349.500	45	**7.767**		
Total	2701.000	54			

Tests of Between-Subjects Effects

Dependent Variable: EYEGLAN

The first *F*-value in the table is an overall *F* test that combines main effects and the interaction in one analysis. This is like the one-way ANOVA you did in Chapter 10. The remaining *F*-values are just like the Excel print-out except that there are no *F*-critical values provided, and the *p* value is now indicated as "Sig." If the "significance" value is less than your alpha level (which is usually 0.05), then you can reject the null hypothesis.

How would these results be reported in a scientific journal article?

If the journal required APA format, the results would be reported in a format something like this:

> There was a significant difference in the number of eye glances based on pair gender (male-male vs. male-female vs. female-female), $F(2, 45) = 16.37, p < .01$. These results suggest that pair gender has a significant effect on the number of eye glances, as can be seen in the figure: Male-male pairs have relatively fewer eye glances, regardless of the discussion type, while male-female and female-female pairs exhibit increasing numbers of eye glances, again regardless of the discussion type.

> There was no significant difference in the number of eye glances due to the effects of discussion type, $F(2, 45) = 1.95, p = .15$.

> There was no significant difference in the number of eye glances based on the interaction between pair gender and discussion type over and above the differences that are due to the main effects of pair gender and discussion type, $F(4, 45) = .43, p = .79$.

These formal sentences include the dependent variable (eye glances), the independent variables (gender pairing, discussion type, and the interaction between gender pairing and discussion type), as well as a statement about statistical significance, the symbol of the test (*F*), the degrees of freedom between groups (2 or 4), the degrees of freedom within groups (45), the statistical values (16.37, 1.95, and .43), and the probability of obtaining these results simply due to chance ($< .01$, .15, and .79, respectively).

When is it appropriate to use a two-way ANOVA test?

1. When you have two independent variables and a between-groups (independent groups) design

2. When the sampling distribution is normally distributed. Again, this is satisfied when

   a. The sample size is greater or equal to 30 in each group ($n \geq 30$) or

   b. The null hypothesis population is known to be normally distributed.

3. When the dependent variable is on an interval or ratio scale (although some statisticians believe that this is too strict)

4. When the variances of the groups are the same or are homogeneous. The homogeneity of variance assumption (HOV) requires that the variances of the samples are equal or, in practical terms, not significantly different from one another. As before, we will use a simple rule of thumb to decide whether the variances are similar enough to be considered homogeneous: If the $\frac{\text{largest variance}}{\text{smallest variance}} \leq 4$, then the HOV assumption is met (okay) and you can proceed with ANOVA. If not, you cannot proceed with the ANOVA (we'll explain what to do instead later).

MULTIFACTORIAL ANOVA

Multifactorial ANOVA is simply the situation where you have even more than two independent variables. Three-way, four-way, and more-way ANOVAs are conceptually possible, although the interaction terms become more and more difficult to interpret. This is because you can have two-way and three-way interactions in the same three-way ANOVA analysis. You will also hear the term *MANOVA,* which describes multiple ANOVA, or the situation where you have multiple *dependent variables,* not independent variables. This topic is beyond the scope of this textbook, but it is important to keep analyses with multiple independent variables conceptually separate from analyses with multiple dependent variables.

Linear or Additive Equation Applied to the Multifactorial ANOVA

There are multiple potential influences (represented by each of your independent variables) on the outcome for each score (or value of your dependent variable), and those influences can be defined with a linear equation. If you are doing a three-way ANOVA, then an individual's score could be represented in the following way:

$$\text{Score} = \mu + IV_1 + IV_2 + IV_3 + (IV_1)(IV_2) + (IV_1)(IV_3) + (IV_2)(IV_3) + (IV_1)(IV_2)(IV_3) + E,$$

where

μ = the grand mean,

IV_1 = effect of the first independent variable,

IV_2 = effect of the second independent variable,

IV_3 = effect of the third independent variable,

$(IV_1)(IV_2)$ = interaction effect of the first two independent variables,

$(IV_1)(IV_3)$ = interaction effect of the first and third independent variables,

$(IV_2)(IV_3)$ = interaction effect of the second and third independent variables,

$(IV_1)(IV_2)(IV_3)$ = interaction effect of all three independent variables,

E = unexplained error.

You can see that these analyses can become quite complex, and interpretation of interactions above two-way interactions is frequently impossible, but the ability to test a large number of hypotheses within a single analytical framework is important. Besides the issues of interpretation, there are also the same issues as with any parametric tests, such as assumptions of normality and homogeneity of variance, all of which can become more cumbersome with more complex designs.

REPEATED-MEASURES ANOVA

A within-groups or repeated-measures ANOVA is similar to a paired *t* test except that you have more than two levels of your independent variable. In other words, you have a within-groups design (subjects are repeated or paired/matched), but your independent variable might include three or more levels. For example, let's say that you want to examine the effect of age-related changes on attention span and you measure the same 50 subjects three times. On the first occasion that you bring them into the laboratory, they are 3 years old, the second time they are 5 years old, and the last time they are 7 years old. Since the same subjects were used in all three "treatments," we can estimate error or variance due to differences among subjects (Subj) alone. Thus, each score is the result of the grand mean (μ) and the added effects of any treatment effect attributable to our (repeated) independent variable (IV = Time), any effect attributable to individual differences between subjects (Subj), and unexplained error (E).

Linear or Additive Equation

$$Score = \mu + IV_1 + Subj + E.$$

In the same way that the paired t test was inherently more powerful than the independent t test, the repeated-measures one-way ANOVA is typically more powerful than the independent one-way ANOVA. This is because we have "removed" or "explained" more of the "unexplained" variance. Specifically, we have removed the portion of the variance that was due to individual or subject differences. We can calculate those differences now because we have multiple measures on the same subjects. It is quite apparent when a subject is consistently above or below the mean, and this variability can be quantified. The remaining unexplained variance (or error) continues to be the number used in the denominator to calculate the F-obtained value. Having a smaller denominator (because we have reduced the unexplained error) results in a larger F-obtained value, making it more likely that we can reject the null hypothesis when it is actually false. We have increased power! Repeated-measures ANOVA is similar to the one-way independent ANOVA from Chapter 10 because you do not calculate an interaction between your independent variable and effect of subject (Subj).

When is it appropriate to use a one-way repeated-measures ANOVA?

1. When you have one independent variable with more than two levels and a within-groups design

2. When the sampling distribution is normally distributed. Again, this is satisfied when

   a. The sample size is greater or equal to 30 in each group ($n \geq 30$) or

   b. The null hypothesis population is known to be normally distributed.

3. When the dependent variable is on an interval or ratio scale (although some statisticians believe that this is too strict)

Notice that the flowchart does not yet tell you what to do if you do not meet the assumptions of your statistical tests. In fact, there are some options when you have three or more samples or two or more factors, but you do not meet the assumptions for parametric tests. One option is to perform a log transformation on your data to determine if the mathematical transformation will result in a normal distribution.

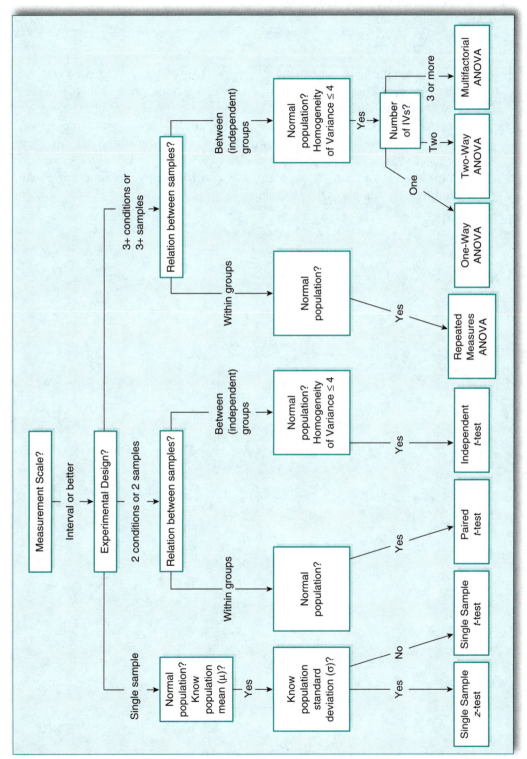

Overview of Single-Sample, Two-Sample, and Three or More Sample Tests: Expanded Flowchart

SUMMARY

In this chapter, we have focused on particular types of ANOVA where you have a within-groups design or a between-groups (independent) design with multiple independent variables. Despite our coverage here, there are many more possibilities, including within-groups designs with more than one independent variable and situations where some of the data that are collected are interval or ratio, but other variables you want to include are not interval or ratio and so on. We'll introduce some other powerful approaches using ANOVA in Chapter 13.

CHAPTER 11 HOMEWORK

Provide a short answer for the following questions.

1. How does a repeated-measures ANOVA differ from a one-way ANOVA?

2. What are the assumptions for a two-way ANOVA?

3. What are the assumptions for a repeated-measures ANOVA?

4. How does a two-way ANOVA differ from a one-way ANOVA?

5. What does the linear equation (presented twice in this chapter) help us to understand?

6. Why is the repeated-measures ANOVA more powerful than the one-way (independent) ANOVA?

7. What is a main effect?

8. What is an interaction effect?

9. Does the repeated-measures ANOVA have an interaction term? Why or why not?

Complete the ANOVA summary tables for Problems 10 to 12.

10. You have just completed a study with three treatment groups (low, medium, and high). There were 10 males and 10 females (or 20 participants) in each of the three treatment groups. Complete the ANOVA source table provided below. Using $\alpha = 0.05$, what do you conclude? Reject or fail to reject the H_0 for each F-obtained value.

Source of Variation	SS	df	MS	F	F crit
Row (sex)	828.8167	1			
Columns (treatment)	1265.633	2			
Interaction (Sex × Treatment)	847.2333	2			
Within	14768.9	54			
Total					

ANOVA

11. You have just been contacted by a colleague who needs help with analyzing a data set. The study had four treatment groups (married, divorced, single, and separated). There were 8 males and 8 females (or 16 participants) in each of the four treatment groups. Complete the ANOVA source table provided below. Using $\alpha = 0.05$, what do you conclude? Reject or fail to reject the H_0 for each F-obtained value.

Source of Variation	SS	df	MS	F	F crit
Row (sex)	495.0625	1			
Columns (status)	1791.625	3			
Interaction (Sex × Status)	1855.063	3			
Within	15426.25	56			
Total					

ANOVA

12. A researcher evaluated the effects of job category and blood pressure on the amount of fines those individuals had paid on parking and traffic tickets in the past 10 years. The two job categories were lawyer and doctor. There were four categories of blood pressure (optimal/below normal, normal, high normal, and very high). They had six lawyers and six doctors in each of the four categories of blood pressure that they evaluated for parking/traffic fines. They hypothesized that individuals with higher blood pressure were more likely to be stressed and thus more likely to be in a hurry and be fined for illegal parking or moving violations. Both job categories were high on income, so the penalty for parking/moving violations would not be financially difficult in and of itself. Complete the ANOVA source table below and interpret the outcome of the study based on the hypothesis and whether or not you can reject the null hypothesis for each F value.

Source of Variation	SS	df	MS	F	F crit
Row (job)	560952.5	1			
Columns (blood pressure)	10028590	3			
Interaction (Job × Blood Pressure)	199493.2	3			
Within	10005058	40			
Total					

ANOVA

Choose the most appropriate and powerful test:

13. Five different elementary schools were selected to participate in a study on reading comprehension. The fifth-grade classes at each school were assessed for reading comprehension using a standardized exam at the beginning of the study, and there was no significant difference found between the groups. Four different reading programs were used for the next 6 months, and one school continued to teach its students using the techniques that had been used previously (control group). After the 6-month period, the reading comprehension scores of fifth graders at the five schools were again compared to one another with the following results:

Treatment A	Treatment B	Treatment C	Treatment D	Control
Mean = 45	Mean = 55	Mean = 40	Mean = 60	Mean = 40
$N = 60$	$N = 50$	$N = 77$	$N = 68$	$N = 54$

Variances between the groups were homogeneous. What is the appropriate analysis for these results? Why is that the correct analysis?

14. Two groups of seniors with Alzheimer's disease were compared on their ability to remember the details of an event that occurred in the previous week. Memory was measured as the number of specific details they were able to remember from the event. If no correct details were remembered, they received a score of zero. The scores ranged from 0 to 19. The first group ($n = 30$) was receiving Drug A, and the second group ($n = 44$) was receiving Drug B. The variances were homogeneous between the two groups. What is the proper analysis to compare the memory capabilities of the two groups based on their drug treatment and why?

15. A cognitive psychologist exploring normal memory decay brought 40 individuals into the laboratory to "plant" a specific memory. This memory was the exact details of the laboratory experience. The 40 individuals were brought back 1 month later, 6 months later, 1 year later, and 3 years later to be assessed on their ability to remember the details of their first experience. Five subjects failed to appear for the last assessment and were dropped from the study and analysis. Memory was recorded as the number of specific (and correct) details recalled about the event. Each error reduced their score by one digit. Scores ranged from –3 to +12. The variances between groups were homogeneous. What is the proper analysis to use to compare their memories at each of the laboratory visits?

Short-answer questions:

16. What is the linear equation for ANOVA?

17. What is the linear equation for two-way ANOVA?

18. What is the linear equation for repeated-measures ANOVA?

The following homework questions should be answered with the online data set provided for this chapter via the textbook's website.

19. Produce a table of descriptive statistics using Microsoft Excel or SPSS.

20. Interpret the descriptive statistics produced by Excel or SPSS. Do you meet the assumption of normality?

21. Do you meet the assumption of homogeneity of variance?

22. Analyze the data set using a two-way ANOVA, and indicate if you have a statistically significant result. Explain how you evaluated the output.

CHAPTER 12

Correlation and Regression

General Overview

Correlation

Pearson's *r* Correlation Coefficient

Linear Regression

Standard Error of the Estimate

Multiple Regression

Effect Sizes and Power

Summary

Chapter 12 Homework

In this chapter, we discuss linear relationships between two variables. For example, we can ask about the relationship between the number of hours that a student studies and his or her overall grade point average. In this case, both the independent and the dependent variables are measured on a noncategorical scale—ordinal, interval, or ratio—and the variables are typically referred to as X and Y. In previous chapters, the independent variable has always been categorical (e.g., male vs. female or low dose, high dose, and control dose), but that is not always the case, and we need to perform the appropriate analyses when our independent variable is continuous rather than categorical. This is also frequently the result of research that involves what some researchers call "quasi-independent" variables: independent variables that are not manipulated or set by the researcher, as in classical experimental designs, but rather are measured at the same time as the result or outcome, the dependent variable. To clarify this distinction, regression designs frequently refer to what we have called "independent variables" as "predictor variables" and refer to "dependent variables" as "criterion variables." An example will help illustrate the concept behind this shift in research approach.

In this chapter, we deviate a bit from the traditional approach to teaching correlation and regression concepts. The reason we do this is twofold. First, we present this chapter with the section on inferential statistics because that allows us to compare analysis of variance (ANOVA) and regression (they are related). We also focus on a conceptual, rather than computational, approach. The mathematics get more cumbersome here, and we believe that the concepts, including what the statistic represents, how to interpret it, and when to use it, are the key lessons. Most statisticians do not calculate these statistics by hand but use a computer software package to produce the results. Thus, understanding how to interpret the results and when to perform these tests are topics that we believe take precedence here. We do provide computational examples in the sidebars for those who are interested in the details.

GENERAL OVERVIEW

Suppose that we examine the relationship between the number of hours a week that 10 students spend on studying and their cumulative grade point average (GPA; Table 12.1). Both variables are continuous. We can graph those results as a scatter plot in Microsoft Excel (Figure 12.1).

Table 12.1

Student	Hours (X)	Grade (Y)
1	10	3.4
2	12	3.5
3	5	2.7
4	2	1.7
5	20	3.9
6	14	3.6
7	11	3.3
8	22	3.8
9	3	1.9
10	18	4.0

Raw Data

The paired data points from all of our 10 students are shown clearly in Figure 12.1. For each student, we went over X units and up by Y units to plot their data. If you do this for all of your data points, you can get a visual estimate of the overall relationship between X and Y. In this case, it appears that as X increases, Y increases. In fact, there appears to be a linear relationship between X and Y. We could draw a straight line that would approximate the relationship between our two variables, essentially by creating a line that is

Figure 12.1

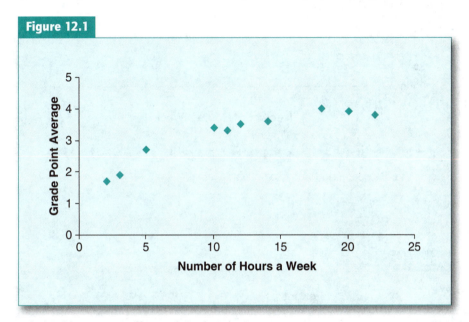

Relationship Between Effort and Grade Point Average

the "average" of all of the data (Figure 12.2). This is a lot like the concept of a grand mean that was introduced in the ANOVA chapter.

In this example, the relationship between your X and Y variables is **positive** (**direct**). The alternative would be a **negative** or **inverse relationship,** such as the one you would expect between the amount of partying and the GPA (higher level of partying is associated with a lower GPA). This example is not an example of a **perfect relationship** because many of the points do not fall exactly on the best-fit line, so it is an **imperfect relationship** (Figure 12.3).

Positive (direct) relationships: As X increases, Y increases.

Negative (inverse) relationships: As X increases, Y decreases.

Perfect relationships: All data points fall on the best-fit line.

Imperfect relationships: All data points do not fall on the best-fit line.

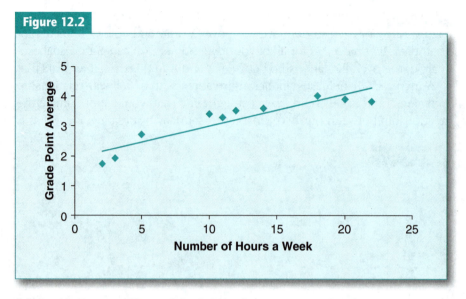

Relationship Between Effort and Grade Point Average

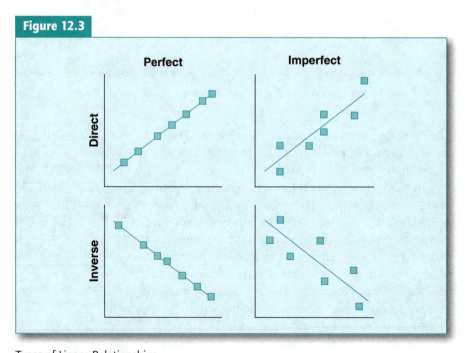

Types of Linear Relationships

CORRELATION

A correlation coefficient is a statistic that expresses the degree of the fit of the data to a line and the type of relationship (direct or inverse). Correlation coefficients make no assumptions about the cause-and-effect direction of an *X-Y* relationship but rather simply measure the degree to which two sets of paired scores vary together in a consistent (linear) manner. So a "significant" correlation between two variables could be produced by (1) a cause-and-effect relationship between *X* and *Y,* (2) a cause-and-effect relationship between *Y* and *X,* or (3) the fact that a third, unmeasured variable is affecting both *X* and *Y,* producing the appearance of a relationship between *X* and *Y!* So be careful when interpreting a correlational relationship. This also implies that you can calculate a correlation coefficient for regression-type "cause-and-effect" data, but you cannot necessarily calculate a "cause-and-effect" regression line for correlational data.

Correlation coefficients have the following characteristics:

1. Values always range between –1 and +1.

2. A positive coefficient indicates a direct relationship (positive slope), whereas a negative coefficient (negative slope) indicates an inverse relationship.

3. A coefficient that is equal to zero indicates that there is no relationship between the two variables (supports your null hypothesis).

4. A coefficient that is equal to –1.00 indicates that you have a perfect inverse relationship between your variables, while a coefficient that is equal to +1.00 indicates that there is a perfect direct (positive) relationship between your variables.

PEARSON'S *r* CORRELATION COEFFICIENT

The most common correlation coefficient is "Pearson's *r*" or "Pearson product-moment correlation coefficient," but there are many other correlation coefficients to choose from. Pearson's *r* requires that your data (*X* and *Y* variables) be interval or ratio scale. The Pearson *r* is a measure of the extent to which your paired scores occupy the same or opposite relative positions in their own distribution. Intuitively, that suggests that lower *X* values are associated with lower *Y* values (direct relationship) or that lower *X* values are associated with higher *Y* values (inverse relationship).

Interpreting the Correlation Coefficient

Pearson's r coefficient for our study time–GPA example data is 0.91. A correlation coefficient this close to perfect (near 1) clearly indicates a strong relationship between our two variables. But what we really need to do is to determine the probability that we could get a statistic of 0.91 or more extreme by chance, the same question that we have been asking throughout many of the earlier chapters. For Pearson's r, we are determining the probability that, given a sample coefficient of 0.91, the correlation coefficient of the underlying population is really zero, reflecting no relationship between X and Y.

FYI: FORMULA FOR PEARSON'S r

$$r = \frac{\sum XY - \frac{(\sum X)(\sum Y)}{n}}{\sqrt{\left[\sum X^2 - \frac{(\sum X)^2}{n}\right]\left[\sum Y^2 - \frac{(\sum Y)^2}{n}\right]}} = \frac{\sum XY - \frac{(\sum X)(\sum Y)}{n}}{\sqrt{(SS_X)(SS_Y)}}$$

Note that the formula requires that you calculate the sums of squares for both X and Y, but the numerator contains a formula that involves the cross products of X and Y that we used in the linear regression calculations. In fact, the numerator of Pearson's r coefficient formula is the same as the numerator of the slope. No wonder the sign of Pearson's r coefficient indicates the direction of the relationship or slope!

Calculating Pearson's r Coefficient for the Effort and GPA Example

 Step 1: Calculate the intermediate numbers required for the formula.

Student	Hours(X)	X^2	Grade(Y)	Y^2	X * Y
1	10	100	3.4	11.56	34.0
2	12	144	3.5	12.25	42.0
3	5	25	2.7	7.29	13.5
4	2	4	1.7	2.89	3.4
5	20	400	3.9	15.21	78.0
6	14	196	3.6	12.96	50.4
7	11	121	3.3	10.89	36.3
8	22	484	3.8	14.44	83.6

Student	Hours(X)	X²	Grade(Y)	Y²	X * Y
9	3	9	1.9	3.61	5.7
10	18	324	4.0	16.00	72.0
Sums	117	1,807	31.8	107.10	418.9

$$SS_X = 1,807 - \frac{(117)^2}{10} = 438.1.$$

$$SS_Y = 107.1 - \frac{(31.8)^2}{10} = 5.976.$$

★ Step 2: Use Pearson's r formula to calculate the coefficient.

$$r = \frac{\sum XY - \frac{(\sum X)(\sum Y)}{n}}{\sqrt{(SS_X)(SS_Y)}}.$$

$$r = \frac{418.9 - \frac{(117)(31.8)}{10}}{\sqrt{(438.1)(5.976)}} = \frac{46.84}{51.167232} = 0.9154296.$$

Basically, our hypothesis test looks very much like a single-sample t test: Our sample value is r, our (hypothesized) population value is zero, and all that is needed is an estimate of the standard deviation of r for the denominator. To make the whole process very simple, all of this work has been done for us. Like our earlier statistical tests, there is a table for critical values of r based on the degrees of freedom $(N - 2)$ and the desired level of alpha (Table E see the Appendix).

Why are the degrees of freedom equal to $N - 2$ (not $N - 1$, as in the t test)? It's because, in this design, we have two continuous variables, and we are estimating variance simultaneously from both measures. Pearson's r is really a measure of the degree to which X and Y covary or vary in the same, or opposite, directions. So we have to estimate variance twice, thus losing two degrees of freedom.

So for our study effort and GPA example, we have an r value of 0.91 and 8 degrees of freedom. If we use an alpha of 0.05, the critical value of r at which we can reject the null hypothesis that this r is not significantly different from 0 is 0.6319; our observed value of r exceeds this critical value, and therefore, we reject the null hypothesis. There is a significant (nonzero) correlation between study time and GPA, and our best estimate of the strength of that correlation is $r = 0.91$.

Even though you have shown a high correlation between the amounts of time spent studying each week and grade point average, you cannot demonstrate a causal link between the two variables with correlational results. It is possible that some third factor that you didn't measure (such as intelligence quotient [IQ]) is really driving both of your variables independently from one another.

Another way to express the relationship between the X and Y variables is to square your r-obtained value to make it R^2.

$$r = \sqrt{\text{proportion of total variance of } Y \text{ that is explained by } X}.$$

$$R^2 = \text{proportion of the variability in } Y \text{ that is explained by } X.$$

The use of the capitalized R is intentional as R^2 is the formal symbol for this statistic. In some uses, r^2 is used for simple correlations between two variables, while R^2 is reserved for the multiple correlations of more complex systems of variables.

If we apply this concept to our example, then R^2 is the proportion of the variance in grade point average that is explained by how much time students spent studying. Obviously, other possible factors influence grade point average, such as how hard you study, your natural aptitude, and the quality of the instruction at your college. If $r = 0.9154296$, then $R^2 = (0.9154296)^2 = 0.838011$. This means that approximately 84% of the variability in grade point average is explained by how much time students spend studying. What if your correlation coefficient is much smaller, so that $r = 0.30$?

$$R^2 = (0.30)(0.30) = 0.09.$$

$$0.09 \times 100 = 9\%.$$

If this were the case, then only 9% of the variability in your Y variable would be explained by your X variable, with 91% of the variability in your Y variable left unexplained. This would suggest that the factor you are studying does not have a strong relationship with your dependent variable.

Results when you use Microsoft Excel to calculate a correlation:

	Hours (X)	Grade (Y)
Hours (X)	1	
Grade (Y)	0.91543	1

Excel gives you a very simple table that contains your X and Y variables and the correlation between hours and grade (0.91543). The ones in the table are the correlation between hours with hours and grade with grade, which is 100%. Note that Excel does not provide probabilities to test the null ($r = 0$) hypothesis.

Results when you use SPSS to calculate a correlation:

		Hours	GPA
Hours	Pearson Correlation	1.000	.915
	Sig. (2-tailed)	.	.000
	N	10	10
GPA	Pearson Correlation	.915	1.000
	Sig. (2-tailed)	.000	.
	N	10	10

Correlations

**Correlation is significant at the 0.01 level (2-tailed).

Unlike Excel, SPSS gives you the correlation as well as the significance (or p value) associated with the correlation. Once again, they have filled in 1.000 when comparing a variable to itself. You can ignore those entries.

How would these results be reported in a scientific journal article?

If the journal required American Psychological Association (APA) format, the results would be reported in a format something like this:

There was a significant correlation between the time spent studying and the grade received, $r = .91, p < .01$.

This formal sentence includes the criterion variable (grade), the predictor variable (time spent studying), as well as a statement about statistical significance, the symbol of the test (r), the statistical value (.91), and the probability of obtaining this result simply due to chance (< .01).

When is it appropriate to use a correlation coefficient?

1. When you have two variables on an interval or ratio (continuous) scale

2. When the relationship between the two variables is linear (rather than curvilinear, or not fitting a straight line)

3. You wish to describe the strength of the relationship between your two variables

At times, the data will consist of one or two ordinal variables—to keep a theme going, say, hours of study and ranking in the class (first, second, third, etc.), or even an ordinal scale of study effort (extensive, moderate, little, none) and class rank. In this case, Pearson's r correlation coefficient is not appropriate, but just as with the t test and ANOVA assumptions, there are alternatives. In this example, Spearman's rank order correlation coefficient is appropriate, and this test is discussed in the nonparametric statistics chapter (Chapter 14).

LINEAR REGRESSION

Linear regression is a technique that is closely related to correlation. In regression, we generally assume that the X variable is the predictor variable (number of hours of study effort, in our example) and the Y variable is the criterion variable (GPA). In other words, we know assume a cause-and-effect relationship between X and Y. This is an important distinction between correlation and regression. It is not a required distinction; that is, you need not automatically assume a cause-and-effect relationship. You may use the regression statistics simply as another descriptive statistic, but it is very common that a cause-and-effect relationship is assumed, and many textbooks "require" it.

In the discussion of our example data and correlation, we described a straight line that was the "best-fit" description of the relationship between X and Y. The correlation coefficient described how well the data fit that line. In regression, the line that we have drawn can be expressed mathematically. Note that our predictor variable (or X variable) is shown on the horizontal or *X*-axis, while our criterion variable (*Y*) is shown on the vertical or *Y*-axis. It is expressed in the following general form:

$Y = bX + a$ (this is the form that Excel has used and is shown in Figure 12.4)

or also commonly reversed:

$$Y = a + bX,$$

where

$X = X$ score;

$Y = Y$ score;

> **Y-intercept:** The value of *Y* when *X* is equal to zero, which is where the line crosses the *Y*-axis.

$a =$ **Y-intercept,** or the value of *Y* when *X* is equal to zero, the point at which the line crosses the *Y*-axis;

> **Slope:** The change in *Y* divided by the change in *X*.

$b =$ **slope** of the line $= \dfrac{\Delta Y}{\Delta X}$ or $\dfrac{\text{rise}}{\text{run}}$ (slope is frequently described as the degree to which a line "rises" [change in the *Y*-axis] for a certain "run" [change in the *X*-axis]).

Figure 12.4

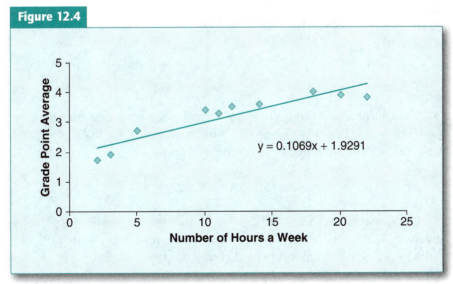

$y = 0.1069x + 1.9291$

Relationship Between Effort and Grade Point Average

Linear regression is a technique that uses the relationship between two variables for prediction. This implies a causal relationship between your two variables. If this is a safe assumption, you can then use the equation of the line to predict Y values for any given X (within the boundaries of your range of X scores). The best fit or "least squares" regression line is the line that minimizes the squared differences between your actual Y and your predicted Y values (Y'). The slope and the Y-intercept are calculated using the least squares method that you learned when we were talking about calculating a mean in Chapter 4. In fact, you can think of the line as the "mean" for each X value. With minor modifications of the general form of the linear equation,

$$Y = bX + a,$$

we can create the least squares regression line:

$$Y' = b_y X + a_y,$$

where

Y' = predicted value of Y,

b_y = slope of the line that minimizes the errors in predicting Y from X,

a_y = Y-intercept of the line that minimizes the errors in predicting Y from X.

Equation of the Line for Our Example Data

You can produce the best-fit line equation for your data directly from Excel, which is shown in Figure 12.2, but, in a sidebar, we'll also show you how to calculate these numbers on your own.

$$Y' = 0.1069(X) + 1.9291.$$

Once you have this formula, you can use it to predict values of Y given an X of your choice. For example, you could predict the GPA of a student who works on homework/studying approximately 13 hours a week.

FYI: CALCULATING THE *Y*-INTERCEPT AND THE SLOPE MANUALLY

Where did Excel come up with the values for the *Y*-intercept and the slope? These can be calculated using the following formulas:

$$b_y = \frac{\sum XY - \frac{(\sum X)(\sum Y)}{n}}{\sum X^2 - \frac{(\sum X)^2}{n}},$$

$$a_y = \bar{Y} - b_y \bar{X},$$

where

   n = number of pairs of scores,

   $\sum XY$ = sum of the product of each *X* and *Y* pair, or the sum of the cross products.

Let's apply this to our previous example to illustrate calculating the *Y*-intercept and the slope without the aid of a computer.

 Step 1: Calculate the intermediate numbers required for the formula.

Student	Hours (X)	X²	Grade (Y)	Y²	X * Y
1	10	100	3.4	11.56	34.0
2	12	144	3.5	12.25	42.0
3	5	25	2.7	7.29	13.5
4	2	4	1.7	2.89	3.4
5	20	400	3.9	15.21	78.0
6	14	196	3.6	12.96	50.4
7	11	121	3.3	10.89	36.3
8	22	484	3.8	14.44	83.6
9	3	9	1.9	3.61	5.7
10	18	324	4.0	16.00	72.0
Sums	117	1,807	31.8	107.1	418.9

$$\bar{Y} = \frac{31.8}{10} = 3.18.$$

$$\bar{X} = \frac{117}{10} = 11.7.$$

(Continued)

(Continued)

 Step 2: Use the formulas to calculate the Y-intercept and slope.

$$b_y = \frac{418.9 - \frac{(117)(31.8)}{10}}{1807 - \frac{(117)^2}{10}} = \frac{46.84}{438.1} = 0.1069.$$

$$a_y = 3.18 - (0.1069)(11.7) = 1.929.$$

 Step 3: Insert the Y-intercept and slope values into the regression line equation.

$$Y' = 0.1069(X) + 1.929.$$

An X value of 13 would yield a Y value of 3.3.

$$Y' = 0.1069(13) + 1.9291 = 3.3188.$$

Based on the information provided in this data set, your best estimate of the GPA of a student who studied 13 hours a week would be 3.3.

Equation of the Line Applied to a New Value of X

$$Y' = 0.1069(X) + 1.9291.$$

$$Y' = 0.1069(13) + 1.9291 = 3.3188.$$

Based on the information provided in this data set, your best estimate of the GPA of a student who studied 13 hours a week would be 3.3.

STANDARD ERROR OF THE ESTIMATE

While we have succeeded in estimating GPA from effort, we need to know the variability around our estimate. If the least squares regression line can be thought of as a "mean," then what is the "standard deviation"? In this case, we use the "standard error of the estimate" (SEE). This is the amount of error around the estimate, just like the standard deviation measures error around the mean.

Because the formula for calculating SEE is cumbersome, yet the concept is similar to what you have already done, we will provide the SEE in the example and exercise problems for you and focus on interpretation instead of manual calculation.

FYI: FORMULA FOR CALCULATING THE STANDARD ERROR OF THE ESTIMATE (SEE)

$$s_{y/x} = \sqrt{\frac{SS_Y - \frac{\left[\sum XY - \frac{(\sum X)(\sum Y)}{n}\right]^2}{SS_X}}{n - 2}}$$

So SEE represents the variability of the data points around the regression line. A very small SEE means that the data points cluster very close to the predicted line, and the degree to which the line accurately represents the individual scores is high. A relatively large SEE means that, while the line is the best representation of the predicted relationship between the X and Y variables, the degree to which the line represents the scores is low.

For the example on studying and GPA, the SEE = 0.35. This means that the variability of the points around the regression line is 0.35 grade point units and allows us to get away from statements such as, "The overall ability to predict GPA from effort spent in studying is good, but not great." Indeed, the larger your SEE, the less confident we can be in any of the predications of Y (Y'). In this case, we could, for example, say that 68% of scores fall within 0.35 grade point units of the predicted value of GPA since we know from our knowledge of z scores that approximately 68% of scores fall within one standard deviation of the mean: In this case, 68% of scores fall within one SEE of the line.

Results if you use Microsoft Excel to calculate a regression on study time–GPA data:

Regression Statistics						
Multiple R	**0.91542963**					
R Square	**0.838011408**					
Adjusted R Square	0.817762834					

(Continued)

(Continued)

Regression Statistics						
Standard Error	**0.347858417**					
Observations	**10**					

ANOVA						
	Df	**SS**	**MS**	**F**	**Significance F**	
Regression	1	5.007956174	5.007956174	41.38619382	0.000201874	
Residual	8	0.968043826	0.121005478			
Total	9	5.976				

	Coefficients	**Standard Error**	**t Stat**	**P-value**	**Lower 95%**	**Upper 95%**
Intercept	**1.929080119**	0.223406068	8.634859975	2.5101E-05	1.413904468	2.44425577
Hours(X)	**0.106916229**	0.016619421	6.433210227	0.000201874	0.068591751	0.145240708

Summary Output

This is the Excel output for the study time and GPA data. We have bolded the numbers that are comparable to the numbers that we just manually calculated or that you were given in the problem (SEE). Note that the F-obtained value (Excel calls it "F") is testing whether we have a significant overall relationship between X and Y. In addition, they have provided you with the exact probability ("P-value") of obtaining an F-obtained value ("F") of 41.38619382, which is 0.000201874. These results suggest that *exactly 0.0201874%* of the time, you would obtain this measure of association between X and Y simply due to chance. Thus, the likelihood that the association is due to chance is not high, but you still do not know whether X causes Y.

Results if you use SPSS to calculate the regression:

Model	R	R Square	Adjusted R Square	Std. Error of the Estimate
1	**.915**	**.838**	.818	**.3479**

Model Summary

a. Predictors: (Constant), HOURS

Model		Sum of Squares	Df	Mean Square	F	Sig.
1	Regression	5.008	1	5.008	41.386	.000
	Residual	.968	8	.121		
	Total	5.976	9			

ANOVA

a. Predictors: (Constant), HOURS

b. Dependent Variable: GPA

Model		Unstandardized Coefficients		Standardized Coefficients	t	Sig.
		B	Std. Error	Beta		
1	(Constant)	**1.929**	.223		8.635	.000
	HOURS	**.107**	.017	.915	6.433	.000

Coefficients

a. Dependent Variable: GPA

The output from SPSS is similar to the output from Excel, but the Y-intercept and the slope are less clearly labeled. They appear in the last table marked "Coefficients" and are highlighted. Note that both Excel and SPSS have given us our r-obtained value along with our regression results.

How would these results be reported in a scientific journal article?

If the journal required APA format, the results would be reported in a format something like this:

There is a significant linear relationship between hours studying and grade earned, such that longer study times are associated with higher grades, $F(1, 8) = 41.39, p < .001; R^2 = .84$.

This formal sentence includes the criterion variable (GPA), the predictor variable (hours studying), as well as a statement about statistical significance, the symbol of the test (*F*), the degrees of freedom (1, 8), the statistical value (41.39), the probability of obtaining this result simply due to chance (< .001), and the proportion of the variance in GPA that appears to be explained by hours studying (.84).

When is it appropriate to use a regression analysis?

1. When you have a predictor and a criterion variable on an interval or ratio scale

2. When the relationship between the two variables is linear (rather than curvilinear, or not fitting a straight line)

3. When your data are **homoscedastic.** This means that the variability around the regression line is uniform for all of the values of *X* (see Figure 12.5). In this figure, the scores in the lower figure show much more variability at higher values of *X.* This is non-homoscedastic or heteroscedastic data, and this breaks a fundamental assumption in interpreting the results of linear regression analyses, just as skewed data alter the interpretation of means and standard deviations.

> Homoscedastic: Variance around line is uniform across all *X*s.

Linear or Additive Equation

Each score (*Y*) is the result of the combined influence of the *X* variable and error (E, represented in linear regression as the SEE). We can add the error component to the equation for the regression line that we developed earlier in the chapter:

$$Y = a + b(X) + \text{SEE}.$$

This looks similar to the linear equation that we developed for ANOVA, and indeed they are related (see Chapter 13).

MULTIPLE REGRESSION

Multiple regression is like the two-way ANOVA because it has more than one predictor variable effect that is assessed at the same time. If we were to add

Figure 12.5

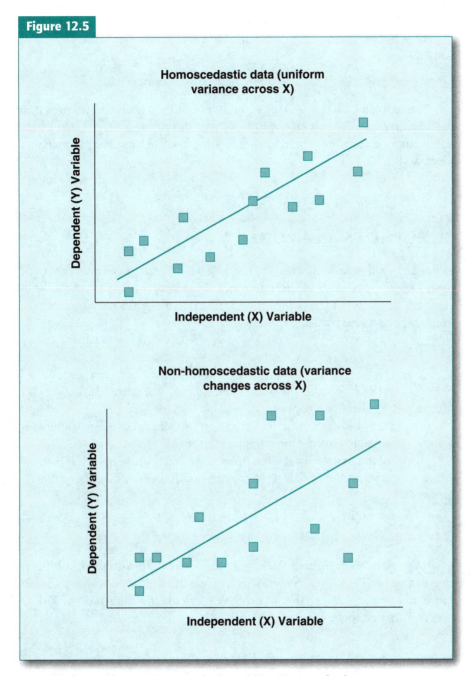

Example of Data That Are Homoscedastic and Non-Homoscedastic

a measure of IQ to our earlier example, we could look at the independent effects of both IQ and effort (studying) on GPA.

Linear or Additive Equation for Multiple Regression

Conceptually, each score (Y) is the result of the combined influence of each X variable and error (E or SEE). We can add the second predictor variable component to the equation for the simple regression line that we just discussed:

$$Y = a + b_1(X_1) + b_2(X_2) + E.$$

EFFECT SIZES AND POWER

Effect size is a standardized measure of the difference between two (or more) group means; it is the difference in means divided by the shared standard deviation of the two or more groups. For regression, you can estimate the effect size by using Cohen's f^2:

$$\text{Cohen's } f^2 = \frac{R^2}{1 - R^2}.$$

Note: R^2 can be obtained from your statistical outputs, particularly if you set the output for the "long" or detailed version.

If you do not know your effect size, you can still proceed by using software and the values from your statistical output.

The program that we have often recommended to students is "G*Power," which can be downloaded at www.psycho.uni-duesseldorf.de/aap/projects/gpower/.

Other Resources for Calculating Power

If you have an experimental design that does not fit with the options for calculating power with G*Power, you can use the following website to find more free resources on the Internet for calculating statistics, including power: http://statpages.org/.

SUMMARY

Linear regression and correlation are closely related, but there are some differences between them to help us distinguish which one is most appropriate for a given situation. Linear regression is used for prediction purposes and calculates an equation for the "best-fit" line that answers the question, "What is the relationship between X and Y?" Correlation analysis focuses on the strength of the relationship between X and Y or how tightly the data points are clustered around the "best-fit" line.

Excel Step-by-Step: Step-by-Step Instructions for Using Microsoft Excel 2003 or 2007 to Run Correlation or Regression Analyses

1. Your first step will be to open Microsoft Excel and type the raw data into a spreadsheet (data listed below). It is helpful to type the column headers so that your output will be labeled later. Note that these are the data from Table 12.1.

Student	Hours (X)	Grade (Y)
1	10	3.4
2	12	3.5
3	5	2.7
4	2	1.7
5	20	3.9
6	14	3.6
7	11	3.3
8	22	3.8
9	3	1.9
10	18	4.0

2. Once your data are entered, you will need to calculate correlation and regression.

For Excel 2003: To find the *t* test you will need, you can go to the built-in data analysis function. You'll find the option under the "Tools" menu, and at the bottom of the list that pops up, you should see "Data Analysis."

If you do not see the "Data Analysis" option under the Tools menu, select "Add Ins" under the Tool menu. Check the box next to "Analysis

TookPak" and click OK. Follow any further instructions that the computer gives you.

For Excel 2007: To get to the data analysis option, click on the "DATA" tab, and the "Data Analysis" tool will be in the "Analysis" section to the far right of the screen. Once you find the Data Analysis tool, the rest of the instructions are the same.

3. After you click on "Data Analysis," a list of possible statistical tests will pop up. Page down the list until you find the appropriate test. We want to do "Correlation" and "Regression." After you have selected the appropriate test, the program will take you through the steps to complete the test. For example, input your range (highlight range of data with your mouse). Note that for regression, the range of each variable (*X* or *Y*) is input separately.

4. Example output from Excel is provided in the chapter for both correlation and regression. Note that Excel does not provide a *p* value with correlation, but SPSS does provide a *p* value.

SPSS Step-by-Step: Step-by-Step Instructions for Using SPSS to Run a Correlation

1. Your first step will be to open SPSS and select the option that allows you to type in new data.

2. This will open a page called "Variable View." To confirm that, look at the tab at the bottom left of the page. There should be two tabs, and one will say "Variable View" (the one you are in now), and the other will say "Data View."

3. Now you need to establish your variables for SPSS. Make a variable name for your *X* variable (HOURS) in the first box of row 1. Now name your *Y* variable (GRADE) and type that into box 1 in row 2. By default, SPSS will consider each of these variables to be numeric, and for these purposes, all of the default codes will work perfectly. However, keep in mind that this is where you can change some of your options to allow for alphabetical data, define coding in your variables, and so on.

4. Click on the "Data View" tab now. You should see that the variable names you entered in Variable View have now appeared at the top of this spreadsheet. Now you can enter your raw data:

Hours (X)	Grade (Y)
10	3.4
12	3.5
5	2.7
2	1.7
20	3.9
14	3.6
11	3.3
22	3.8
3	1.9
18	4.0

1. From the SPSS menu, you should now select "Analyze," then "Correlation," and finally "Bivariate." This will open up a new pop-up window with your variables listed on the left-hand side. Select your X variable and use the arrow and move it to the box as your "X variable." Now move your Y variable the same way over to the "Y" box.

2. Choose the "Pearson" test and hit "OK" to run the test. You'll get the output for the correlation value and the significance or p value.

SPSS Step-by-Step: Step-by-Step Instructions for Using SPSS to Run a Regression

1. Your first step will be to open SPSS and select the option that allows you to type in new data.

2. This will open a page called "Variable View." To confirm that, look at the tab at the bottom left of the page. There should be two tabs, and one will say "Variable View" (the one you are in now), and the other will say "Data View."

3. Now you need to establish your variables for SPSS. Make a variable name for your X variable (HOURS) in the first box of row 1. Now name your Y variable (GRADE) and type that into box 1 in row 2. By default, SPSS will consider each of these variables to be numeric, and for these purposes, all of the default codes will work perfectly. However, keep in mind that this is where you can change some of your options to allow for alphabetical data, define coding in your variables, and so on.

4. Click on the "Data View" tab now. You should see that the variable names you entered in Variable View have now appeared at the top of this spreadsheet. Now you can enter your raw data:

Hours (X)	Grade (Y)
10	3.4
12	3.5
5	2.7
2	1.7
20	3.9
14	3.6
11	3.3
22	3.8
3	1.9
18	4.0

5. From the SPSS menu, you should now select "Analyze," then "Regression," and finally "Linear." This will open up a new pop-up window with your variables listed on the left-hand side. Select your X variable and use the arrow and move it to the box as your "X variable." Now move your Y variable the same way over to the "Y" box.

6. Click "OK" to run the test. You'll get the output for the list of variables, a model summary, R, R2, adjusted R2, standard error of the estimate, and an ANOVA table. Refer to Chapter 13 to understand why an F-table would appear with your output on a regression test.

CHAPTER 12 HOMEWORK

Provide a short answer for the following questions.

1. What is R^2 (or the coefficient of determination) and how do you calculate it?

2. What is R^2 (or the coefficient of determination) for a correlation of $r = 0.89$?

3. How would you put the answer to Question 2 into words (i.e., what does it tell us about the variability of X and Y)?

4. What is the possible range for a correlation coefficient?

5. What two considerations must you take into account when deciding which correlation coefficient to use?

6. What assumption does Pearson's r make about the relationship between X and Y?

7. Researchers are interested in whether the number of children in a family predicts the number of times that family eats out at restaurants. They randomly sample 10 volunteers from a local city; the experimenters measure the number of children they currently have at home (variable X) and the number of days a year they eat out at restaurants (variable Y). For the data below, calculate the appropriate correlation between the two variables.

$$\Sigma X = 19$$

$$\Sigma X^2 = 71$$

$$\Sigma XY = 1,782$$

$$\Sigma Y = 1,035$$

$$\Sigma Y^2 = 108,293$$

8. Find the least squares regression line for the data in Question 7.

9. Correlation and regression differ in that _____.

   A. correlation is primarily concerned with the size and direction of relationships

   B. regression is primarily used for prediction

   C. both a and b are true

   D. neither a nor b are true

10. In an imperfect relationship, _____.

   A. all the points fall on the line

   B. a relationship exists, but all of the points do not fall on the line

C. no relationship exists

D. a relationship exists, but none of the points can fall on the line

11. The closer the points on a scatter diagram fall to the regression line, the _____ between the scores.

A. higher the correlation

B. lower the correlation

C. correlation doesn't change

D. Need more information

12. The lowest degree of correlation shown below is _____.

A. 0.75

B. −0.33

C. −0.25

D. 0.15

13. Knowing nothing more than that IQ and memory scores are correlated 0.84, you could validly conclude that _____.

A. good memory causes high IQ

B. high IQ causes good memory

C. neither good memory nor high IQ cause each other

D. a third variable causes both good memory and high IQ

E. None of the above

14. A correlation between college entrance exam grades and scholastic achievement was found to be −1.08. On the basis of this, you would tell the university that:

A. The entrance exam is a good predictor of success

B. They should fire that sorry statistician and hire a new one

C. The exam is a poor predictor of success

D. Students who do best on this exam will make the worst students

E. Students at this school are underachieving

15. An experimenter believes that reaction time on a word identification task might be related to one's score on a vocabulary test. She tested a random

sample of people from a normal population. Reaction time was measured in milliseconds, and the score on a vocabulary test was out of 100 points.

$$\Sigma X = 5,290$$

$$\Sigma X^2 = 2,935,750$$

$$\Sigma XY = 420,070$$

$$\Sigma Y = 804$$

$$\Sigma Y^2 = 65,382$$

$$N = 10$$

$$Sy/x = 8.22$$

A. Calculate the appropriate correlation coefficient.

B. Calculate the slope of the regression line.

C. Calculate the Y-intercept of the regression line.

D. Given the reaction time of 400 ms, predict this person's score on the vocabulary test.

E. Based on the information you have, how confident are you about your prediction?

Choose the most appropriate and powerful test:

16. A sports psychologist analyzed the batting average of subjects that hit right-handed ($n = 59$), subjects that hit left-handed ($n = 41$), and subjects that hit from either side (switch hitters! [$n = 30$]). The subjects were randomly selected from active major league players across the American and National Leagues. The variances are 0.18, 0.24, and 0.32, respectively.

17. Research on reaction time indicates that, on average, people press a button after a bell goes off in 1.88 seconds. A random sample of 16 subjects from a normally distributed population of airline pilots was assessed using the reaction time test to determine if the reaction time of pilots is significantly faster than the reaction time in the general population. Their mean reaction time was 1.61 seconds with a standard deviation of 0.07 seconds.

18. Over a 50-year period, the number of children per couple in China has been 3.88, with a standard deviation of 0.91. Recently, China has tried to decrease the number of children born per couple. A sample of 30 couples living in China

was performed, and they had an average of 1.38 children with a standard deviation of 0.25. Have the new policies resulted in a significant change in the number of children per family?

19. A researcher is interested to know whether there is any difference in spatial aptitude for air traffic controllers and artists. Thirty-five air traffic controllers and 35 artists were randomly selected to participate in the study. Both groups were given a spatial skills test in which the number of questions answered correctly was evaluated. Air traffic controllers averaged a mean of 23 problems right with a standard deviation of 4. Artists averaged a mean of 27 problems right with a standard deviation of 6. Can the researcher conclude that air traffic controllers and artists differ in their spatial aptitude?

20. A researcher performed a study examining the impact divorce has on anxiety. She randomly sampled 101 married individuals, 31 of whom later became divorced. To measure anxiety, she recorded heartbeats per minute of the 31 divorced individuals before and after they were divorced. How should she analyze the data?

Short-answer questions:

21. What is the standard linear equation of the least squares regression line?

22. What does it mean when your data are homoscedastic?

23. What does it mean when your data are heteroscedastic?

The following homework questions should be answered with the online data set provided for this chapter via the textbook's website.

24. Analyze the data set using either Excel or SPSS. Produce a correlation and indicate if you have a statistically significant result. Explain how you evaluated the output.

25. Analyze the data set using either Excel or SPSS. Produce a regression analysis and indicate if you have a statistically significant result. Explain how you evaluated the output.

CHAPTER 13

General Linear Model

Linear or Additive Equation/Model

Application of the General Linear Model Approach

Analysis of Covariance (ANCOVA)

Summary

Chapter 13 Homework

T he general linear model is a relatively new (past 20–30 years) approach to parametric statistics. Statisticians have been rethinking both the use and the interpretation of the common statistical tests, such as analysis of variance (ANOVA), analysis of covariance (ANCOVA), regression, multiple regression, and t tests, and have realized that all of these statistics have a common mathematical underpinning. In fact, the general linear model underlies even some sophisticated statistical techniques that are typically taught in modern graduate-level courses, such as factor analysis, cluster analysis, multidimensional scaling, and discriminate function analysis. We have been foreshadowing this common underpinning in several of the previous chapters of the book. Here we will formally describe this relationship and why it is useful to us. Some of our graduate teaching assistants have been astounded and enlightened by reading this chapter of our textbook. This is not surprising given that even most graduate-level statistics courses are broken up into separate courses for "ANOVA" and "Regression." Professors may "hand wave" and say that the two are related, but this is rarely emphasized or explained.

Because this is an introductory statistics textbook, this chapter is meant to be an overview or introduction rather than a complete treatment of the subject. To keep it conceptual and improve the flow of the chapter, we have not included the usual number of calculated examples, and there are no other extra sections. Our goal is to convince you that the underlying mathematical model for ANOVA, regression, and t tests is the same. We feel that this chapter is important for students of statistics because it will help you learn how to run analyses on a computer (how to set up your model) and how to interpret your results. In fact, some statistical packages will give you an F value when you run a regression! Read further to find the computer output from SPSS.

LINEAR OR ADDITIVE EQUATION/MODEL

This concept was first discussed in Chapter 10 (ANOVA) where we said that each score is the result of a combined influence of the grand mean (μ), any treatment effect attributable to our independent variable (IV), and unexplained error (E).

Conceptually, this states that there are multiple potential influences on the outcome for each score, and those influences can be defined with a linear equation in which each influence is added together. We did not present this information in Chapter 9 ("Two-Sample Tests"), but both conceptually and mathematically, ANOVA is the same as a t test but simply with more groups. In fact, the F statistic that is calculated in an ANOVA with only two groups is simply t^2 for an independent t test with the same two groups. If you accept that ANOVA is actually a t test with more groups, then it probably won't surprise you to find out that their linear equations are the same.

ANOVA/t Test

$$\text{Score} = \mu + \text{IV} + \text{E}.$$

The formula for regression is really the same as the formula for ANOVA. The notation looks a bit different, but that is because each test was derived independently of the other test, and historically they have each had their own terminology and notation.

Regression

$$Y = a + bX + E,$$

where

$X = X$ score,

$Y = Y$ score,

$a = Y$-intercept,

$b = $ slope of the line,

$E = $ unexplained error,

Slope $(b) = \dfrac{\Delta Y}{\Delta X}$ or $\dfrac{\text{rise}}{\text{run}}$.

So, the grand mean is the same as the y-intercept, the effect of the IV is reflected in a slope across X and unexplained, and leftover variance is E or error variance. ANOVA, t tests, and regression are the same! Not only are t tests, one-way ANOVA, and linear regression related to one another, but complex ANOVA and multiple regression designs are also variations of this linear equation or the "general linear model."

Two-Way ANOVA

If there are two independent variables that could potentially influence each score, then you can define the effect of each independent variable, as well as the interactive effect of the two variables, conceptually with the linear equation:

$$\text{Score} = \mu + IV_1 + IV_2 + (IV_1)(IV_2) + E,$$

where

$\mu = $ combined influence of the grand mean,

$IV_1 = $ effect of the first independent variable,

$IV_2 = $ effect of the second independent variable,

$(IV_1)(IV_2) = $ interaction effect of the independent variables,

$E = $ unexplained error.

Three-Way ANOVA

If you are measuring three independent variables (a three-way ANOVA), then an individual's score could be represented in the following way:

$$\text{Score} = \mu + IV_1 + IV_2 + IV_3 + (IV_1)(IV_2) + (IV_1)(IV_3) + (IV_2)(IV_3) + (IV_1)(IV_2)(IV_3) + E,$$

where

μ = combined influence of the grand mean,

IV_1 = effect of the first independent variable,

IV_2 = effect of the second independent variable,

IV_3 = effect of the third independent variable,

$(IV_1)(IV_2)$ = interaction effect of the first two independent variables,

$(IV_1)(IV_3)$ = interaction effect of the first and third independent variables,

$(IV_2)(IV_3)$ = interaction effect of the second and third independent variables,

$(IV_1)(IV_2)(IV_3)$ = interaction effect of all three independent variables,

E = unexplained error.

Repeated-Measures ANOVA

Another variation of the ANOVA is the repeated-measures ANOVA where subjects serve in more than one condition of the experiment, or subjects are matched so that they are treated (statistically) as if they are the same individual.

$$\text{Score} = \mu + IV_1 + \text{Subj} + E,$$

where

μ = combined influence of the grand mean,

IV_1 = effect of the independent variable,

Subj = effect attributable to individual differences between subjects,

E = unexplained error.

Multiple Regression

Multiple regression is like simple regression but with an additional continuous independent variable. It is the regression parallel to two-way ANOVA.

$$Y = a + b_1 X_1 + b_2 X_2 + E,$$

where

$X_1 = X$ score for the first independent variable,

$X_2 = X$ score for the second independent variable,

$Y = Y$ score,

$a = Y$-intercept,

$b_1 = $ slope of the first line,

$b_2 = $ slope of the second line,

$E = $ unexplained error.

Comparing Two-Way ANOVA to Multiple Regression

The slope of the lines in regression represents the main effects in ANOVA. If the lines are separated, there is a main effect. If the lines cross or have significantly different slopes, then you have an interaction. To represent this visually, we return to the graphs shown to represent interactions in Chapter 11. Graph "a" represents a significant main effect for sex (males vs. females) but no interaction between the factor of sex and the group factor (we know this because the slopes of the lines are parallel).

Assumptions of the General Linear Model

All of these formulas share in common similar assumptions (i.e., normal distribution) and a linear form where the effects of each variable on the score are additive. Thus, all of these equations are fundamentally the same form of linear equation or model, and they can all be solved using the same process (matrix algebra).

APPLICATION OF THE GENERAL LINEAR MODEL APPROACH

What advantages can we gain by conceptually understanding that all of these tests are based on the general linear model (GLM)? We can gain two very practical advantages.

1. *You can understand and interpret why there is an* F-*obtained value in your linear regression output.* Modern statistical software will automatically give you an F-obtained value when you conduct a regression analysis. The F value indicates how well the slope and intercept of your regression line fit your data. It is an overall analysis of how well this line fits your data, much like the ANOVA is an overall test of more than two means. Likewise, you will also often see t-obtained values on the output. These t values are like multiple comparisons are for one-way ANOVA because they are testing how well individual parts (like your slope or your Y-intercept) fit your

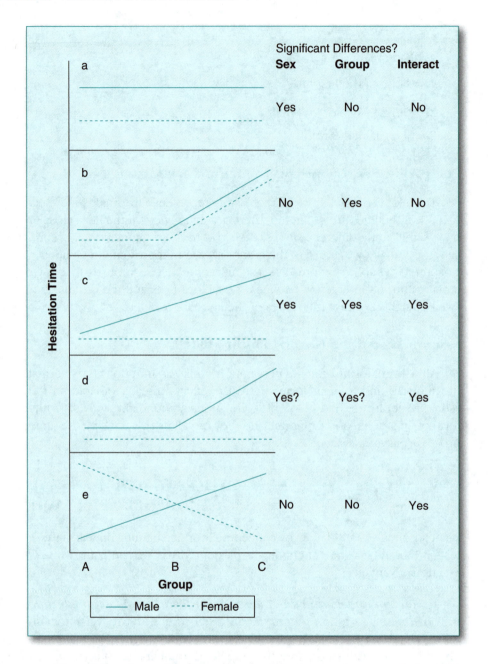

overall model, and you do not need to look at them unless your *F* obtained (or overall test) is significant.

2. *You can mix and match the measurement scales of your independent variables.* Until this point, we have kept ANOVA and regression fairly

separate in our discussions, stating that ANOVA was suited to situations where you could group your independent variable into nominal categories (sex, age groups, drug dosages, etc.) and that regression was appropriate when your independent variable was continuous (or interval/ratio). However, you might come up with an experimental design with more than two independent variables where one of them is nominal and one of them is continuous. In this case, neither the classic ANOVA design nor a regression design is appropriate, yet an overall test is still desired. You might wish to compare the effects of an antidepressant medication based on age (interval scale) and dose (broken into control, low-dose, and high-dose groups). Using a subject's actual age rather than grouping subjects by broad age ranges (e.g., < 20, 20–40, 40–60, > 60) is more powerful statistically because you are not collapsing your more powerful interval/ratio data into less powerful ordinal categories.

3. *You can address the problem of predictor variables that are correlated with one another.* Complex designs (with multiple factors) allow you to assess pure main effects of particular variables or factors, after you have removed the correlation with other factors. For example, if sex and weight are correlated, but only the variable of weight is important for interpreting the effects of a drug treatment, then the model will address the intercorrelation of sex and body weight and give you the remaining main effect of weight after the contribution of sex is removed.

SPSS Example

A clinical psychologist tests the effects of three levels of an antidepressant medication on the number of days that subjects report feeling depressed. The subject's age is included as an additional independent variable of interest as there have been some reports that treatment can be influenced by the age of the subject, and the researcher is particularly interested in any interactions between dose and age. Table 13.1 contains the raw data entered into SPSS, and the subsequent tables are abbreviated versions of the SPSS output from the analysis. The analysis was run in SPSS Version 10.0 for Windows using the "general linear model" command (see Figure 13.1).

The initial steps to conduct a two-way ANOVA in SPSS were as follows:

Select "Analyze" from the toolbar

Select "General Linear Model"

Select "Univariate"

Figure 13.1

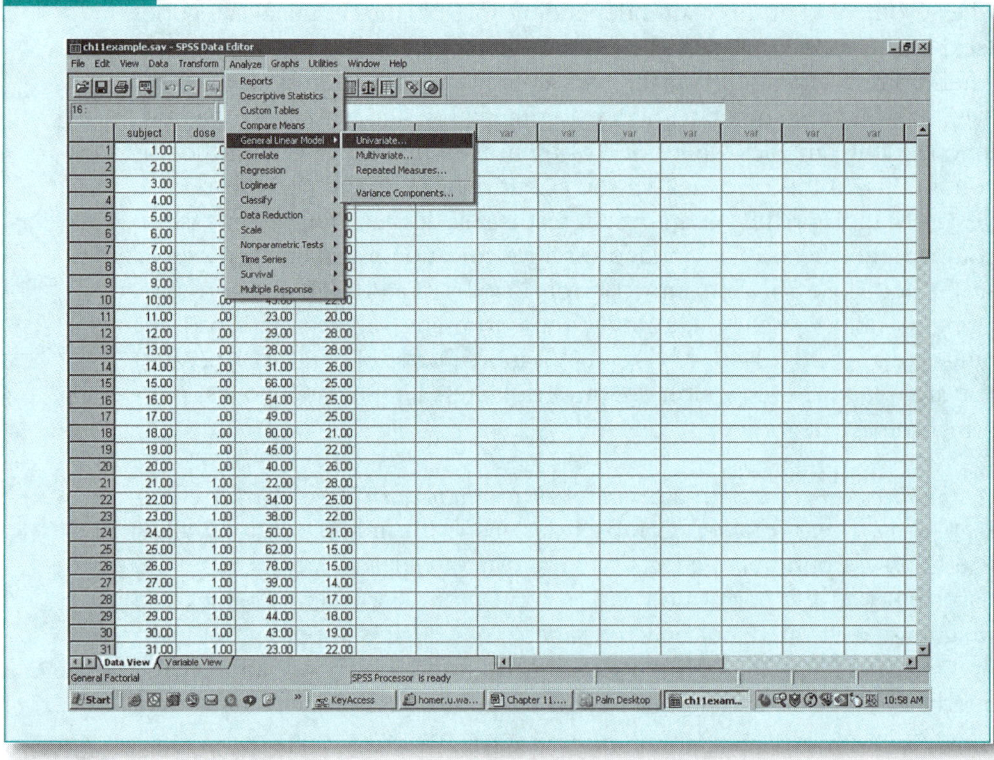

Choosing the General Linear Model Option in SPSS

SPSS will use the default model of a "Full-factorial" design. This is what you want because it includes both of the main effects and the interaction term. You can also click the box to include a test of the *y*-intercept in the analysis (traditionally, the most valuable test is the text of an IV effect—the slope—but you can also test for whether the grand mean/*y*-intercept differs significantly from zero). The output has been slightly condensed from SPSS.

The main effects for dose and age are indicated in the table, and dose is significant (Sig. value is less than .05 alpha value), but age is not significant. There is no significant interaction between dose and age (DOSE * AGE). The "Corrected Model" is an overall test of the fit of your model to the data. Note that it is not significant, most likely because both one of your main effects and your interaction are not statistically significant. These results suggest that the number of days feeling depressed varies with the dose of the anti-depressant drug, but there is no relationship between age and feeling

Table 13.1

Source	Type III Sum of Squares	df	Mean Square	F	Sig.
Corrected Model	4005.000	70	57.214	1.599	.228
Intercept	18992.757	1	18992.757	530.853	.000
DOSE	**1165.244**	**2**	**582.622**	**16.284**	**.001**
AGE	**1063.032**	**29**	**36.656**	**1.025**	**.520**
DOSE * AGE	**862.310**	**39**	**22.111**	**.618**	**.856**
Error	**322.000**	**9**	**35.778**		
Total	26772.000	80			
Corrected Total	4327.000	79			

Dependent Variable: DAYS

Tests of Between-Subjects Effects

a. R Squared = .926 (Adjusted R Squared = .347)

depressed and no interaction between dose and age (i.e., the effect of dose is the same across all ages, or the effect of age [none] is the same at all three dose levels). We have successfully analyzed a data set using GLM where one independent variable was continuous (age) and would typically have been explored using traditional regression techniques and where the second independent variable was nominal (dose) and would have traditionally been analyzed with ANOVA.

This flexibility and the opportunity to assess the subsequent interaction between these two independent variables on different scales are the primary advantages to the GLM technique.

ANALYSIS OF COVARIANCE (ANCOVA)

Analysis of covariance, or ANCOVA, is a specialized form of ANOVA that is used when an investigator wishes to remove the effects of a variable that is known to influence the dependent variable but is not the subject of the current experiment and analysis. For example, we might have added body weight to the previous example because doses can react differently in individuals of vastly different body weights, but that was not the topic of interest to the clinical psychologists. If they had added body weight into the analysis, it would be considered a covariate. They would do so in an attempt to remove that unexplained variance from the error term. This would increase their ability to detect any main effects or interactions of their independent variables. ANCOVA has been replaced by GLM. By adding an additional variable, you can achieve the desired effect of partitioning out (removing) the variance associated with it before continuing on to perform the analysis of the variables of interest. SPSS has a separate category into which you can place these covariate variables. This is advantageous because the covariate(s) will automatically be included first, before the primary variables, and the covariate(s) will be excluded from interaction terms, in which you are often not interested when it comes to covariates.

SUMMARY

The general linear modeling approach is very powerful, allowing the simultaneous assessment of continuous and categorical independent variables.

The concept of general linear modeling has been widely embraced by the research community but is still rarely included in undergraduate statistics textbooks. More important for this course, these concepts help us to interpret the results of statistical software packages that use the GLM terminology, which has become universal.

Provide a short answer for the following questions.

1. What are two practical advantages to using a GLM approach?

2. What does ANCOVA stand for?

3. What is the purpose of ANCOVA?

4. How are ANOVA, regression, and *t* tests similar to one another?

5. Why haven't we always known about GLM, and why aren't the terms in all of the formulas in Question 4 all the same?

6. Why is this linear approach also known as an additive model?

7. If all of these equations are fundamentally the same form of linear equation or model, how can they all be solved?

8. What categories of independent variables can be used with a general linear model?

   A. Categorical

   B. Continuous

   C. Categorical and continuous

   D. None of the above

CHAPTER 14
Nonparametric Tests

W e've told you all about analyzing data if you meet the assumptions of the indicated tests, but what the heck are you supposed to do if you don't meet the assumptions? You are probably wondering about that by now. It's time to give you some options.

Parametric tests, such as t tests, z tests, and F tests, require that populations are normally distributed and, in the case of the independent t test and analysis of variance (ANOVA), that the variances are equal. Traditional nonparametric tests do not have these assumptions but are much *less powerful* than parametric tests because they are based on converting your data to ranks. This has the effect of converting information-rich interval/ratio data to less informative ordinal data. Thus, you should use parametric tests whenever possible. But when an assumption is extremely violated or the data are from the nominal or ordinal measurement scale, then you should use nonparametric tests or consider alternative tests with other assumptions, such an unequal variances test (e.g., Welch's *t* test) or resampling methods (discussed in Chapter 15).

Despite being less powerful than parametric tests, nonparametric tests are useful and are relatively easy to calculate. We will begin with the appropriate test to use when your data are from the nominal measurement scale and then turn to tests that use the ordinal scale. Keep in mind that these tests are appropriate either when your data are originally collected on these scales or when your data on an interval or ratio scale violate the assumptions of more powerful tests and therefore must be converted to an ordinal scale for these tests. Keep in mind that nonparametric tests are extremely useful because they are flexible, are easy to calculate, and may be most appropriate for your data set. Many of these tests have direct parametric equivalent tests, so you will recognize the experimental designs in the following examples. This should help you assimilate the large number of tests we present in this chapter.

CHI-SQUARE

The chi-square test is appropriate when you have nominal independent variable(s) but only group membership (frequencies) as your raw data. For example, you might want to determine whether students in this class are randomly distributed in their political party affiliations (see Table 14.1). Conceptually, this is an appropriate analysis for a chi-square test because the purpose of a chi-square analysis is to compare the *observed* frequencies to the *expected* frequencies. The *expected* frequencies are determined by assuming that group membership is random. If membership in a group differs from random, then we can say that there are statistically significant differences between the actual data and the results that one would expect due to chance (or random factors).

Table 14.1

Democrat	Green	Independent	Republican
40	20	12	40

Number of Students Who Belong to Each of the Four Political Parties (Observed Frequencies)

$N = 40 + 20 + 12 + 40 = 112$ students in the course.

Calculation of Expected Values

If the expected frequencies are based purely on chance, then they are easy to calculate based on how many categories you have in your analysis. In the political party example above, we know that we have four categories of political parties, and the probability of being in each category simply due to chance would be ¼ or 25%. This is the random model of calculating *expected* frequencies.

There is a second form of expected frequencies, called a "goodness-of-fit" model. In this method, you calculate expected frequencies based on some prior knowledge of what those values should be that are not necessarily random (the goodness-of-fit model). For example, you might wish to determine whether the political party frequencies you observe now are significantly different from what they were in the past, so you want to compare today's frequencies (observed frequencies) with the frequencies from, say, 30 years ago and use the historical percentages to calculate your *expected* frequencies. This technique is used frequently in introductory biology or genetics courses where basic Mendelian genetics places traits into nominal categories (like dominant or recessive). For example, it is known that ¾ (75%) of the normal population has the ability to curl their tongue, while ¼ (25%) typically cannot curl their tongue (Table 14.2). This is a dominant inherited trait.

Table 14.2

Can Curl Tongue	Cannot Curl Tongue
75%	25%

Expected Frequencies of Tongue Curling Based on Mendelian Genetics

Clearly this trait is not expected to be inherited based on the random model, which would predict that the ability or inability to curl your tongue would have expected frequencies of 50%, respectively. This is a clear case where the goodness-of-fit model of calculating *expected* frequencies is more appropriate. We will return to both the random and the goodness-of-fit models of calculating expected frequencies when we present two complete examples of chi-square tests, but first we turn to the formula for calculating a chi-square test.

Calculation of the Chi-Square (χ^2) Test

f_e = expected frequency with random sampling from the null hypothesis population,

f_o = observed frequency in the sample,

k = number of cells (or categories).

$$\chi^2 \text{ obtained} = \sum_{cell=1}^{cell=k} \left[\frac{(f_o - f_e)^2}{f_e} \right].$$

The formula requires that you perform the calculations indicated in the brackets for each cell (or category) and then sum the numbers you calculated for each cell (or category) to produce the chi-square (χ^2) obtained value. Once that is done, you must compare the obtained value to a critical value in the same manner that you have done for the parametric tests.

What are you really doing in this calculation? It's very simple:

1. You are calculating a difference score that measures the deviation of the observed counts from the corresponding expected counts (remember the paired t test and difference scores?).

2. You are then squaring the difference score (remember why we had to square the deviations to calculate the standard deviation? Because otherwise they would always sum to zero!).

3. You are then dividing the squared deviation by the expected frequency to standardize the result for the size of the sample.

4. You are calculating this standardized squared deviation score for each category and summing them together to obtain χ^2.

If χ^2 obtained $\geq \chi^2$ critical, then you reject the null hypothesis. If the obtained value is not greater or equal to the critical value, then you fail to reject the null hypothesis. A table of critical values for χ^2 can be found at the back of the book (Table F). Critical values for χ^2 are based on the degrees of freedom ($k - 1$) and the alpha level. The degrees of freedom are similar to the commonly used $N - 1$ (or $n - 1$) for parametric tests, but in this, the "sample size" is really the number of categories, k. Evaluations of χ^2 obtained are always nondirectional.

Complete Example

A sociology professor asked the students in her course to indicate their political party affiliation in an anonymous survey. She found that 40 of her students considered themselves Democrats, 20 were Green Party members, 12 supported the Independents, and 40 were Republicans out of her 112 total students in the course. Is party affiliation distributed randomly across the four parties represented by her class?

Null hypothesis (H_0): Political party affiliations are distributed randomly.

Alternative hypothesis (H_A): Political party affiliations are not randomly distributed.

 Step 1: Determine the *expected* frequency (f_e).

Based on the story problem, it is clear that we are supposed to compare the actual party affiliations of this class with a random model. The f_e for this example would be ¼ or 25% (1 divided by the number of cells) because there are four possible cells, and if the results were random, then one would expect that each cell (or category) is equally likely.

Cell 1	Cell 2	Cell 3	Cell 4
25%	25%	25%	25%

To convert these percentages to a frequency, we need to multiply the percentage by the total *N*.

$$f_e = 112 \times 0.25 = 28.$$

Thus, we have the same expected frequency for each cell.

 Step 2: Calculate the χ^2 obtained.

	Democrat	Green	Independent	Republican
f_o	40	20	12	40
f_e	28	28	28	28

$$\chi^2 \text{ obtained} = \sum_{cell-1}^{cell=k} \left[\frac{(f_o - f_e)^2}{f_e} \right].$$

$$\chi^2 \text{ obtained} = \left[\frac{(40-28)^2}{28} \right] + \left[\frac{(20-28)^2}{28} \right] + \left[\frac{(12-28)^2}{28} \right] + \left[\frac{(40-28)^2}{28} \right].$$

$$\chi^2 \text{ obtained} = \left[\frac{(12)^2}{28} \right] + \left[\frac{(-8)^2}{28} \right] + \left[\frac{(-16)^2}{28} \right] + \left[\frac{(12)^2}{28} \right].$$

$$\chi^2 \text{ obtained} = \left[\frac{144}{28}\right] + \left[\frac{64}{28}\right] + \left[\frac{256}{28}\right] + \left[\frac{144}{28}\right].$$

$$\chi^2 \text{ obtained} = 9.142857 + 5.142857 + 5.142857 + 2.285714 = 21.71429.$$

★ **Step 3:** Evaluate the probability of obtaining this χ^2 value due to chance.

To evaluate your χ^2-obtained value, you must use the χ^2 distribution (see the Appendix), and to do so, you need an alpha level (0.05) and degrees of freedom ($k - 1$). Thus, the *df* for this problem are $4 - 1 = 3$. Compare the χ^2-critical value with your χ^2-obtained value, and reject the null hypothesis when your χ^2-obtained value is equal to or greater than your χ^2-critical value. When $\alpha = 0.05$ and the degrees of freedom are equal to 3, you should use 7.82 as your χ^2-critical value.

$21.71 \geq 7.82$, so we *reject* the null hypothesis.

How should we interpret these data in light of the randomness of party affiliation for the students in this hypothetical course?

These results suggest that you would expect to obtain these frequencies in party affiliation less than 5% of the time, assuming party affiliation was random. Thus, it is not very likely that these frequencies all come from the random null hypothesis population. Of course, as always, there is some finite probability that you could get frequencies that deviate from the expected frequencies this much simply due to chance, but that probability is less than 5%.

Results of chi-square test using SPSS:

	Observed N	Expected N	Residual
1.00	40	28.0	12.0
2.00	20	28.0	−8.0
3.00	12	28.0	−16.0
4.00	40	28.0	12.0
Total	112		

NPar Tests

Chi-Square Test

Party

	Party
Chi-Square	**21.714**
df	**3**
Asymp. Sig.	.000

Test Statistics

a. 0 cells (.0%) have expected frequencies less than 5. The minimum expected cell frequency is 28.0.

We have highlighted the numbers that are comparable to our manual calculations. The exact probability (SPSS calls it "Sig.") is less than our alpha level of 0.05, and thus we must assume that these results would be *unlikely* simply due to chance. SPSS does not give you a critical value to compare with your obtained value, but it is not necessary since SPSS does give you the exact probability of obtaining these results or results more extreme than these simply due to chance.

How would these results be reported in a scientific journal article?

If the journal required American Psychological Association (APA) format, the results would be reported in a format something like this:

Political party affiliations differ significantly from random, χ^2 (3, $N = 112$) = 21.71, $p < .01$.

This formal sentence includes the independent variable (party affiliation), as well as a statement about statistical significance, the symbol of the test (χ^2), the degrees of freedom (3), the statistical value (21.71), and the estimated probability of obtaining this result simply due to chance ($< .01$).

Second Example (f_e Calculated With the Goodness-of-Fit Model)

The same statistics professor now asks the students in her course to indicate whether or not they can curl their tongues. She finds that 80 of her students can curl their tongues, while 32 of her 112 total students cannot curl their tongues even with practice. Does this sample of students fit the Mendelian pattern that a single gene dominant trait is exhibited by 75% of individuals while the recessive trait is exhibited by 25% of all individuals? Remember that we only have to show that there is no significant difference between our sample and the goodness-of-fit model.

Null hypothesis (H$_o$): Tongue curling in this sample fits the expected model of 75% and 25%.

Alternative hypothesis (H$_A$): This sample does not come from the normal population where 75% of all individuals can curl their tongues and 25% of all individuals cannot curl their tongues.

 Step 1: Determine the expected frequency (f_e).

Based on the story problem, it is clear that we are supposed to compare the actual frequencies of tongue curling for this class with a goodness-of-fit model. The f_e for this example would be determined by converting each percentage to a frequency. We do this by multiplying each percentage by the total N.

$$f_e = 112 \times 0.75 = 84.$$

$$f_e = 112 \times 0.25 = 28.$$

Thus, we have different expected frequencies for each cell.

 Step 2: Calculate the χ^2 obtained.

	Curl	No Curl
f_o	80	32
f_e	84	28

$$\chi^2 \text{ obtained} = \sum_{cell=1}^{cell=k} \left[\frac{(f_o - f_e)^2}{f_e} \right].$$

$$\chi^2 \text{ obtained} = \left[\frac{(80 - 84)^2}{84} \right] + \left[\frac{(32 - 28)^2}{28} \right].$$

$$\chi^2 \text{ obtained} = \left[\frac{(-4)^2}{84} \right] + \left[\frac{(4)^2}{28} \right].$$

$$\chi^2 \text{ obtained} = \left[\frac{16}{84} \right] + \left[\frac{16}{28} \right].$$

$$\chi^2 \text{ obtained} = 0.190476 + 0.571429 = 0.761905.$$

Step 3: Evaluate the probability of obtaining this χ^2 value due to chance.

To evaluate your χ^2-obtained value, you must use the χ^2 distribution (see Table F in the Appendix), and to do so, you need an alpha level (0.05) and degrees of freedom ($k - 1$). Thus, the *df* for this problem are $2 - 1 = 1$. Compare the χ^2-critical value with your χ^2-obtained value, and reject the null hypothesis when your χ^2-obtained value is equal to or greater than your χ^2-critical value. When $\alpha = 0.05$ and the degrees of freedom are equal to 1, you should use 3.84 as your χ^2-critical value.

0.761905 < 3.84, so we *fail to reject* the null hypothesis.

How should we interpret these data?

These results suggest that you would expect to obtain these frequencies in tongue curling more than 5% of the time. Therefore, it is likely that these frequencies all come from the null hypothesis population where 75% of all individuals can curl their tongues and 25% cannot curl their tongues. Our sample appears to come from a population that fits the Mendelian characteristics of a single gene trait that is dominant.

How would these results be reported in a scientific journal article?

If the journal required APA format, the results would be reported in a format something like this:

Tongue curling in this sample is not significantly different than the expected model of 75% and 25%, χ^2 (1, $N = 112$) = .76, $p > .05$.

This formal sentence includes the independent variable (ability to curl or not to curl your tongue), as well as a statement about statistical significance, the symbol of the test (χ^2), the degrees of freedom (1), the statistical value (.76), and the estimated probability of obtaining this result simply due to chance (< .01).

TEST OF INDEPENDENCE BETWEEN TWO VARIABLES: CONTINGENCY TABLE ANALYSIS

The contingency table analysis is a variation of the chi-square test that allows you to test the independence between two variables. Thus, you can ask whether two categorical variables are independent from one another or related to one another. To perform this analysis, the variables must be classified into mutually exclusive categories. As we showed in the earlier chi-square discussion, the cells contain the frequencies. What is a mutually exclusive category? Mutually exclusive categories are designed so that being a member of one category prohibits your inclusion in the other category. For example, consider biological sex as a mutually exclusive category. Being male prohibits one from being in the other category (female). If one of the independent variables is called Factor A and the second one is Factor B, then we can state the null and alternative hypotheses in the following way:

Null hypothesis (H_0): Factor A is not related to Factor B.

Alternative hypothesis (H_A): Factor A and Factor B are related.

Performing a Contingency Table Analysis

The equation for the contingency table analysis is the same one that we used for the chi-square, where you calculate a value for each cell that is determined by the observed and the expected frequencies. However, the f_e and the *df* are calculated differently for this analysis. The expected frequencies are slightly more challenging to calculate because you must do it for each cell, and the value is influenced by the subtotals in the rows and columns. Specifically, the f_e for each cell is calculated in the following way:

$$\frac{(\text{Row marginal for that cell})(\text{Column marginal for that cell})}{\text{Total } N}.$$

Every value or cell in the two-way table is part of a row and a column. The total frequencies in each row are called the row marginals, and the total frequencies for the columns are called the column marginals. So each cell has a corresponding row marginal value and a column marginal value. These are multiplied together, and the result is divided by the total sample size to obtain the expected value for each cell. We'll turn to an example to demonstrate the calculations and the marginal values.

Example

Hallmark Card Company hires a statistician to determine whether or not males and females differ in their preference for a "mushy" Valentine's Day card. They obtain the following results from their survey of 400 adults.

	Love 'em	Hate 'em	Don't Care	Marginals
Males	100	50	50	
Females	150	25	25	
Marginals				

Note that the marginals are calculated by adding up the rows and columns. For example, the first row marginal for males is 200 (100 + 50 + 50), the first column marginal for "love 'em" is 250 (100 + 150), and the total or overall N is 400.

Null hypothesis (H_0): Preference for "mushy" Valentine's Day cards is randomly distributed between males and females.

Alternative hypothesis (H_A): Preference for "mushy" Valentine's Day cards is not randomly distributed between males and females.

 Step 1: Determine the expected frequency (f_e).

Based on the story problem, it is clear that we are supposed to compare the actual frequencies of preferences for mushy Valentine's Day cards with the expected frequencies based on a random model. The expected frequencies will be based on row and column marginals, and there may be different expected frequencies for each cell.
Determine each row and column marginal by adding up the frequencies in the respective row or column.

	Love 'em	Hate 'em	Don't Care	Marginals
Males	Cell 1 100	Cell 2 50	Cell 3 50	200
Females	Cell 4 150	Cell 5 25	Cell 6 25	200
Marginals	250	75	75	400

Note that cells are numbered from left to right (marginals do not count).

Calculate Frequency Expected Values

For each cell f_e = (its row marginal)(its column marginal)
Total N

$$\text{Cell 1 } f_e = \frac{(200)(250)}{400} = 125.$$

$$\text{Cell } 2\, f_e = \frac{(200)(75)}{400} = 37.5.$$

$$\text{Cell } 3\, f_e = \frac{(200)(75)}{400} = 37.5.$$

$$\text{Cell } 4\, f_e = \frac{(200)(250)}{400} = 125.$$

$$\text{Cell } 5\, f_e = \frac{(200)(75)}{400} = 37.5.$$

$$\text{Cell } 6\, f_e = \frac{(200)(75)}{400} = 37.5.$$

 Step 2: Calculate the χ^2 obtained.

f_e values for each cell are indicated in the parentheses.

	Love 'em	Hate 'em	Don't Care	Marginals
Males	100 (125)	50 (37.5)	50 (37.5)	200
Females	150 (125)	25 (37.5)	25 (37.5)	200
Marginals	250	75	75	400

$$\chi^2 \text{ obtained} = \sum_{cell=1}^{cell=k} \left[\frac{(f_o - f_e)^2}{f_e} \right].$$

$$\chi^2 \text{ obtained} = \left[\frac{(100 - 125)^2}{125} \right] + \left[\frac{(50 - 37.5)^2}{37.5} \right] + \left[\frac{(50 - 37.5)^2}{37.5} \right] + \left[\frac{(150 - 125)^2}{125} \right] +$$

$$\left[\frac{(25 - 37.5)^2}{37.5} \right] + \left[\frac{(25 - 37.5)^2}{37.5} \right] =$$

$$\chi^2 \text{ obtained} = \left[\frac{(-25)^2}{125} \right] + \left[\frac{(12.5)^2}{37.5} \right] + \left[\frac{(12.5)^2}{37.5} \right] + \left[\frac{(25)^2}{125} \right] + \left[\frac{(-12.5)^2}{37.5} \right] + \left[\frac{(-12.5)^2}{37.5} \right] =$$

$$\chi^2 \text{ obtained} = \left[\frac{625}{125} \right] + \left[\frac{156.25}{37.5} \right] + \left[\frac{156.25}{37.5} \right] + \left[\frac{625}{125} \right] + \left[\frac{156.25}{37.5} \right] + \left[\frac{156.25}{37.5} \right] =$$

$$\chi^2 \text{ obtained} = 5 + 4.1666 + 4.1666 + 5 + 4.1666 + 4.1666 = 26.6664.$$

★ **Step 3:** Evaluate the probability of obtaining this χ^2 value due to chance.

To evaluate your χ^2-obtained value, you must use the χ^2 distribution (see Table F in the Appendix), and to do so, you need an alpha level (0.05) and degrees of freedom [$df = (r - 1)(c - 1)$], where r = the number of rows and c = the number of columns. Thus, the df for this problem are $(2 - 1)(3 - 1) = 2$. Compare the χ^2-critical value with your χ^2-obtained value, and reject the null hypothesis when your χ^2-obtained value is equal to or greater than your χ^2-critical value. When $\alpha = 0.05$ and the degrees of freedom are equal to 2, you should use 5.99 as your χ^2-critical value. $26.67 > 5.99$, so we *reject* the null hypothesis.

How should we interpret these data?

These results suggest that you would expect to obtain these frequencies in Valentine's Day card preference less than 5% of the time. Thus, it is unlikely that these frequencies come from the null hypothesis population where card preferences are distributed randomly by sex. It appears that sex and preference for mushy Valentine's Day cards are related. In fact, it appears that women tend to prefer mushy valentines (!).

How would these results be reported in a scientific journal article?

If the journal required APA format, the results would be reported in a format something like this:

There was a significant relationship between sex and a preference for mushy Valentine's Day cards, χ^2 (2, $N = 400$) = 26.67, $p < .01$. It appears that women tend to prefer mushy valentines.

This formal sentence includes both independent variables (card preference as well as sex), a statement about statistical significance, the symbol of the test (χ^2), the degrees of freedom (2), the statistical value (26.67), and the estimated probability of obtaining this result simply due to chance ($< .01$).

ASSUMPTIONS FOR THE CHI-SQUARE AND CONTINGENCY TABLE TESTS (χ^2)

There are two assumptions for the chi-square test:

1. Each subject has only one entry, and the categories are mutually exclusive. This ensures that there is independence between your observations.

2. In chi-square designs that are larger than 2×2, the expected frequency in each cell must be at least five. There is recent evidence to suggest that this assumption is unnecessary, or at least that chi-square is not seriously affected by violations of this assumption.

The chi-square test is frequently used with nominal data, but you can convert other scales into mutually exclusive categories and frequencies to employ the chi-square test.

NONPARAMETRIC TESTS TO PERFORM WHEN THE PARAMETRIC TEST CANNOT BE PERFORMED DUE TO VIOLATION OF THE ASSUMPTIONS

Alternative tests can be conducted when one or more of the assumptions of a parametric test are not met. These alternatives do not require a normal distribution or homogeneity of variance or interval/ratio data (although they still require an assumption of random sampling). These tests are typically performed when the assumptions are not met or when the data collected are only on an ordinal scale. However, they can be conducted on data that are originally collected on interval/ratio scale because they convert the data to an ordinal scale as part of the calculations (see Table 14.3).

Table 14.3	
Parametric Test	**Equivalent Nonparametric Test**
Correlated (paired) *t* test	Wilcoxon
Independent *t* test	Mann-Whitney *U*
One-way ANOVA	Kruskal-Wallis

Nonparametric Equivalents of Popular Parametric Tests

WILCOXON SIGNED RANK TEST (*T*)

The Wilcoxon signed rank test (*T*) replaces the correlated (or paired) *t* test. This test considers both the magnitude and the direction of the effect, but it is not as powerful as the *t* test. This test is appropriate when you have a nominal independent variable with two groups (i.e., before vs. after), but your dependent variable is *at least* on an ordinal scale.

Calculation Steps

1. Calculate the differences between the two conditions (e.g., before vs. after). This is your difference score; the Wilcoxon test is similar to the paired t test at this point.

2. Rank the absolute values of the difference scores from the smallest to the largest. If there are ties on the ranked scores, then you should take an average rank, but do not reuse ranks. This will be clearer when you see an example.

3. Assign the sign of the original difference score to the rank (put the signs back on). So if your difference score was negative, you should add that negative back on to the ranked score so that you have a negative rank.

4. Sum all of the positive ranks together and then separately sum the negative ranks. Calculate the absolute value of the summed ranks for each group.

5. The smaller of the summed ranks becomes the T-obtained value.

6. Evaluate with the Wilcoxon table.

Note: Reject if $T_{obt} < T_{crit}$.

Example

A new program is developed to enrich the kindergarten experience of children in preparation for first grade. Compare the test scores of nine children before and after the program is initiated. Note that a score of 10 indicates that the child is performing at the grade level. Compare the pretest and posttest scores for these children.

Pupil	Pretest	Posttest
A	9	16
B	7	12
C	14	13
D	12	10
E	8	14
F	7	11

Pupil	Pretest	Posttest
G	12	15
H	8	10
I	12	14

 Step 1: Calculate the difference scores.

Note that since we are subtracting from left to right, increased scores on the posttest result in negative difference scores (suggesting improvement in the scores). You could also subtract in the opposite direction. As long as you are consistent, it won't affect the result.

Pupil	Pretest	Posttest	Difference
A	9	16	−7
B	7	12	−5
C	14	13	1
D	12	10	2
E	8	14	−6
F	7	11	−4
G	12	15	−3
H	8	10	−2
I	12	14	−2

 Step 2: Rank the absolute value of the difference score.

Organize the raw difference scores from smallest to largest and assign the *absolute value* of each score a rank from left to right in the following way (see note on tied ranks):

Raw Scores	1	2	−2	−2	−3	−4	−5	−6	−7

Raw Scores Only

Raw Scores	1	2	−2	−2	−3	−4	−5	−6	−7
		(2)	(3)	(4)					
Rank	1	3	3	3	5	6	7	8	9

Raw Scores With Ranks Below Them

Why did we use a rank of "3" three times? If you take the absolute value of –2, then you find that you have three scores with an absolute value of 2. The way to handle this is to take the average of the ranks that each of the tied scores would have if they were not tied. In this case, each of these 2s holds the ranks of 2, 3, and 4. If you average these ranks $\frac{(2 + 3 + 4)}{3}$, then the average rank is 3. We use 3 for each of these, but we cannot "reuse" the ranks (2-3-4) again because these scores have already used these ranks. Instead, 5 should be the next rank.

 Step 3: Assign the ranks the original sign of the difference score.

Raw Scores	1	2	–2	–2	–3	–4	–5	–6	–7
Rank	1	3	–3	–3	–5	–6	–7	–8	–9

 Step 4: Sum the ranks for the positive and negative ranks separately.

Sum ranks for positive ranks:

$$1 + 3 = 4.$$

Sum ranks for negative ranks:

$$|–3| + |–3| + |–5| + |–6| + |–7| + |–8| + |–9| = 41.$$

 Step 5: Determine the T-obtained value.

Since the smaller of the two sums is the T-obtained value, in this case, $T_{obt} = 4$.

★ **Step 6:** Evaluate the T-obtained value.

Use the T table for the Wilcoxon (Table G in the Appendix) to evaluate your T-obtained value. The critical value of T is based on your paired sample size ($n = 9$) and your alpha level ($\alpha = 0.05$, two-tailed). Based on this, the T-critical value for this problem is 6. Note that the decision rule is that you can only reject the null hypothesis when $T_{obt} \le T_{crit}$. This is backwards compared to previous tests. In this example, $4 < 6$, so we should *reject* the null hypothesis. Let's look at the SPSS output for these data that we have analyzed by hand.

Results if you use SPSS to calculate the Wilcoxon:

		N	Mean Rank	Sum of Ranks
POSTTEST PRETEST	Negative Ranks	2	2.00	**4.00**
	Positive Ranks	7	5.86	**41.00**
	Ties	0		
	Total	9		

NPar Tests

Wilcoxon Signed Ranks Test

Ranks

a. POSTTEST < PRETEST

b. POSTTEST > PRETEST

c. PRETEST = POSTTEST

	Posttest – Pretest
Z	–2.199
Asymp. Sig. (2-tailed)	**.028**

Test Statistics

a. Based on negative ranks

b. Wilcoxon Signed Ranks Test

We have placed the numbers that are comparable to our manual calculations in bold. The exact probability (SPSS calls it "Sig.") is less than our alpha level of 0.05, and thus we must assume that these results would be *unlikely* simply due to chance. SPSS does not give you a critical value to compare with your obtained value, but it is not necessary since SPSS does give you the exact probability of obtaining these results or results more extreme than these simply due to chance.

How would these results be reported in a scientific journal article?

If the journal required APA format, the results would be reported in a format something like this:

There was a significant improvement between the pre- and posttest scores of children after experiencing the enrichment program, $T(N = 9) = 5.00, p = .03$.

This formal sentence includes the dependent variable (test scores), the independent variable (before vs. after), as well as a statement about statistical significance, the symbol of the test (T), the sample size (9), the statistical value (5.00), and the estimated probability of obtaining this result simply due to chance (.03).

Assumptions for the Wilcoxon Test

1. Raw scores must be at least ordinal but can be interval/ratio.

2. Must have random sampling.

MANN-WHITNEY U TEST (U)

The Mann-Whitney U test is appropriate when you have an independent (or between-groups) design with two groups (e.g., experimental vs. control groups). This test is designed to evaluate the separation between the groups. A larger separation suggests that the populations are unlikely to be the same, while a smaller separation suggests that both groups come from the same underlying null hypothesis population. This test replaces the independent t test when the original data collection is ordinal or when the assumptions of the independent t test are not met.

Calculation Steps

1. Place scores in rank order and rank scores from both samples regardless of experimental condition. Handle tie scores by assigning the average rank as you did with the Wilcoxon test.

2. Sum the ranks separately by experimental group (R_1 and R_2).

3. Calculate two U-obtained values by using the following equations:

$$U \text{ obtained} = n_1 n_2 + \frac{n_1(n_1 + 1)}{2} - R_1.$$

$$U \text{ obtained} = n_1 n_2 + \frac{n_2(n_2 + 1)}{2} - R_2.$$

The smaller value becomes the true U-obtained value while the larger value is referred to as the U' (prime) obtained value.

4. Evaluate the *U*-obtained value using the Mann-Whitney table (in the Appendix) and reject the null hypothesis if $U_{obt} < U_{crit}$.

Complete Example 1

Does prenatal exposure to alcohol reduce SAT scores? Evaluate the SAT scores of one group of teenagers who were exposed to alcohol during prenatal development and another group of teenagers who were not exposed to alcohol in prenatal development.

Control (CONT)	Experimental (EXP)
610	340
600	410
585	365
625	315

Control Group vs. Experimental Group (Prenatal Exposure to Alcohol)

$$n_1 = 4.$$

$$n_2 = 4.$$

★ **Step 1:** Place scores in rank order and rank scores from both samples regardless of experimental condition.

Group	Exp.	Exp.	Exp.	Exp.	Cont.	Cont.	Cont.	Cont.
RawScores	315	340	365	410	585	600	610	625
Rank	1	2	3	4	5	6	7	8

★ **Step 2:** Sum the ranks separately by experimental group (R_1 and R_2).

$$R_1 \text{ (Experimental Group)} = 1 + 2 + 3 + 4 = 10.$$

$$R_2 \text{ (Control Group)} = 5 + 6 + 7 + 8 = 26.$$

★ **Step 3:** Calculate two *U*-obtained values by using the equations

$$U \text{ obtained} = n_1 n_2 + \frac{n_1(n_1 + 1)}{2} - R_1 = 4(4) + \frac{4(4 + 1)}{2} - 10 \text{ and}$$

$$U \text{ obtained} = n_1 n_2 + \frac{n_2(n_2 + 1)}{2} - R_2 = 4(4) + \frac{4(4 + 1)}{2} - 26.$$

Remember that the smaller value becomes the true U-obtained value, while the larger value is referred to as the U' (prime) obtained value:

$$U \text{ obtained} = 16 + 10 - 10 = 16,$$

$$\text{so } U' \text{ obtained} = 16.$$

$$U \text{ obtained} = 16 + 10 - 26 = 0,$$

$$\text{so } U' \text{obtained} = 0.$$

★ **Step 4:** Check your work!

$$U_{obt} + U'_{obt} \text{ should} = (n_1)(n_2)$$

$$0 + 16 = 4(4) \text{ or } 16 = 16$$

Since the two sides of the equation match, you know that you've done your work correctly.

★ **Step 5:** Evaluate the U-obtained value using the Mann-Whitney table (Table H in the Appendix).

In order to look up the U-critical value, you need to know the sample sizes for each group, your α level, and whether or not you are making a one-tailed or a two-tailed test of the hypothesis. Because you are asking whether prenatal exposure to alcohol *reduces* SAT scores, you are conducting a one-tailed test. Under these conditions, your U-critical value is 1. You can also get the U'-critical value, which is 15. We'll show you the decision rules for evaluating either the U obtained or the U' obtained because some students prefer to use one of these rules over the other, but you should get the same conclusion regardless of which number you evaluate.

Decision rule: If U obtained $\leq U$ critical, then reject the null hypothesis.

Decision rule: If U' obtained $\geq U'$ critical, then reject the null hypothesis.

If we use the U-obtained and U-critical values, then $0 \leq 1$, so we can reject the null hypothesis. If we use the U'-obtained and U'-critical values, then $16 \geq 15$, so we can reject the null hypothesis. We get the same conclusion (reject the null hypothesis) regardless of which value we choose to evaluate.

Results if you use SPSS to calculate the Mann-Whitney *U* test:

	Group	N	Mean Rank	Sum of Ranks
Score	1.00	4	6.50	**26.00**
	2.00	4	2.50	**10.00**
	Total	8		

NPar Tests

Mann-Whitney Test

Ranks

	Score
Mann-Whitney U	.000
Wilcoxon W	10.000
Z	−2.309
Asymp. Sig. (2-tailed)	.021
Exact Sig. [2*(1-tailed Sig.)]	.029

Test Statistics

a. Not corrected for ties

b. Grouping Variable: GROUP

The first SPSS table gives you the R_1 and R_2 sums that you calculated by hand as well as the sample size in each group, while the second SPSS table indicates the *U*-obtained value (0) and that the probability of obtaining this value if chance alone is operating is only 0.029.

How would these results be reported in a scientific journal article?

If the journal required APA format, the results would be reported in a format something like this:

There was a significant difference between the SAT scores of one group of teenagers who were exposed to alcohol during prenatal development and another group that was not exposed to alcohol,

$U (N = 8) = 0.00$, $p = .03$. Teenagers who had been exposed to alcohol during prenatal development scored less than teenagers without that exposure.

This formal sentence includes the dependent variable (SAT scores), the independent variable (alcohol exposure vs. control), as well as a statement about statistical significance, the symbol of the test (U), the sample size (8), the statistical value (0), and the estimated probability of obtaining this result simply due to chance (.03).

Assumptions for the Mann-Whitney U Test

1. The dependent variable must be measured in at least the ordinal scale, but interval or ratio data can be "collapsed" into ordinal data via the ranking.

2. It is only appropriate when you have an independent or between-groups design and two groups/conditions.

3. Random sampling is required.

Complete Example 2

A marketing researcher asks whether the desire to purchase Coca-Cola after viewing a commercial varies based on the gender of the viewer. He asks the participants of the study to rate their preference for Coca-Cola after viewing a commercial about Coca-Cola and compares the ratings of the males and females. Because of the specific content of the commercial, he believes that it will have more appeal for females than it will have for males. He predicts that males will be less influenced by the commercial than females and will give lower ratings compared to females.

Males (M)	Females (F)
5	10
6	10
4	8
7	5
5	8
	9
	9

Ratings of Coca-Cola by Males and Females After Viewing a Commercial

$n_1 = 5$

$n_2 = 7$

Step 1: Place scores in rank order and rank scores from both samples regardless of experimental condition.

Ties on ranks—if two numbers are tied:

1. Get an average rank.

2. Don't reuse ranks.

Group		M	M	M	F	M	M	F	F	F	F	F	F
RawScores		4	5	5	5	6	7	8	8	9	9	10	10
		(2)	(3)	(4)			(7)	(8)	(9)	(10)	(11)	(12)	
Rank		1	3	3	3	5	6	7.5	7.5	9.5	9.5	11.5	11.5

Average ranks used to calculate assigned rank:

$$\frac{2 + 3 + 4}{3} = 3.$$

$$\frac{7 + 8}{2} = 7.5.$$

$$\frac{9 + 10}{2} = 9.5.$$

$$\frac{11 + 12}{2} = 11.5.$$

 Step 2: Sum the ranks separately by experimental group (R_1 and R_2).

$$R_1 \text{ (Males)} = 1 + 3 + 3 + 5 + 6 = 18.$$

$$R_2 \text{ (Females)} = 3 + 7.5 + 7.5 + 9.5 + 9.5 + 11.5 + 11.5 = 60.$$

Step 3: Calculate two U-obtained values by using the equations:

$$U \text{ obtained} = n_1 n_2 + \frac{n_1(n_1 + 1)}{2} - R_1 = 5(7) + \frac{5(5 + 1)}{2} - 18.$$

$$U \text{ obtained} = n_1 n_2 + \frac{n_2(n_2 + 1)}{2} - R_2 = 5(7) + \frac{7(7 + 1)}{2} - 60.$$

Remember that the smaller value becomes the true U-obtained value while the larger value is referred to as the U'-obtained value:

$$U \text{ obtained} = 35 + 15 - 18 = 32,$$

$$\text{so } U' \text{ obtained} = 32.$$

$$U \text{ obtained} = 35 + 28 - 60 = 3,$$

$$\text{so } U' \text{ obtained} = 3.$$

Step 4: Check your work!

$$U_{obt} + U'_{obt} \text{ should} = (n_1)(n_2)$$

$$32 + 3 = 5(7) \text{ or } 35 = 35$$

Since the two sides of the equation match, you know that your work is done correctly.

 Step 5: Evaluate the U-obtained value using the Mann-Whitney table (Table H in the Appendix).

Because you are asking whether males are less affected by the commercial than females, you are conducting a one-tailed test. Under these conditions, your U-critical value is 6. You can also get the U'-critical value, which is 29.

Decision rule: If U obtained $\leq U$ critical, then reject the null hypothesis.

Decision rule: If U' obtained $\geq U'$ critical, then reject the null hypothesis.

If we use the U-obtained and U-critical values, then $3 \leq 6$, so we can reject the null hypothesis. If we use the U'-obtained and U'-critical values, then $32 \geq 29$, so we can reject the null hypothesis. We get the same conclusion (reject the null hypothesis) regardless of which value we choose to evaluate.

How would these results be reported in a scientific journal article?

If the journal required APA format, the results would be reported in a format something like this:

Males were significantly less influenced by the commercials than females and gave lower ratings compared to females, U ($N = 12$) = 3.00, $p < .05$.

This formal sentence includes the dependent variable (ranking), the independent variable (sex), as well as a statement about statistical significance, the symbol of the test (U), the sample size (12), the statistical value (3.00), and the estimated probability of obtaining this result simply due to chance ($< .05$).

KRUSKAL-WALLIS TEST (H)

The Kruskal-Wallis test is used when data are in three or more groups and is ordinal, and it replaces the one-way (independent) ANOVA when either or both of the assumptions are broken. Unlike the ANOVA, the Kruskal-Wallis test does not require that the sampling distribution be normally distributed or that the variances between the groups are homogeneous.

Calculation Steps

1. Rank all scores from the smallest to the largest.

    a. Disregard groups initially when ranking.

    b. Handle tied ranks as you did with the previous nonparametric tests.

2. Sum ranks for each group (R_i).

3. Calculate the H-obtained value using the following formula:

$$H_{obt} = \left[\frac{12}{N(N+1)} \right] \left[\sum \frac{(R_i)^2}{n_i} \right] - 3(N+1).$$

4. Evaluate using chi-square table, $df = k - 1$, where k equals the number of groups in the experiment. Note: The decision rule returns to the typical pattern where you reject the null hypothesis if and only if $H_{obt} \geq H_{crit}$

5. If the null hypothesis is rejected, pairwise combinations of Mann-Whitney U tests should be used to determine the post hoc pattern of differences among the groups, just as a series of Tukey tests is used following null hypothesis rejection in ANOVA.

Complete Example

A college instructor asks whether the performance of her students varies with the time of day the class is held. She compares the grade point averages (GPA) of the same course topic held during three different time periods to determine if there is a difference in performance based on time of day alone.

Early a.m.	Early p.m.	Late p.m.
2.5	2.8	3.8
3.1	3.2	3.2
2.2	3.0	4.0
2.6	2.7	3.7
2.2	2.3	3.7

GPA Scores for One Course Held at Three Different Time Periods

★ **Step 1:** Rank scores regardless of group/condition and handle ties appropriately.

Raw scores

2.2	2.2	2.3	2.5	2.6	2.7	2.8	3.0	3.1	3.2	3.2	3.7	3.7	3.8	4.0

(1) (2)   (10) (11) (12) (13)

Rank

1.5 1.5 3 4 5 6 7 8 9 10.5 10.5 12.5 12.5 14 15

Average of ranks used to calculate absolute ranks:

$$\frac{1 + 2}{2} = 1.5.$$

$$\frac{10 + 11}{2} = 10.5.$$

$$\frac{12 + 13}{2} = 12.5.$$

★ **Step 2:** Sum the ranks separately by experimental group.

Early a.m.	(R_1)	Early p.m.	(R_2)	Late p.m.	(R_3)
2.5	4	2.8	7	3.8	14
3.1	9	3.2	10.5	3.2	10.5
2.2	1.5	3.0	8	4.0	15
2.6	5	2.7	6	3.7	12.5
2.2	1.5	2.3	3	3.7	12.5

Sum ranks for each group (R_i):

$\Sigma R_1 = 4 + 9 + 1.5 + 5 + 1.5 = 21.$

$\Sigma (R_1)^2 = 21^2 = 441.$

$\Sigma R_2 = 7 + 10.5 + 8 + 6 + 3 = 34.5.$

$\Sigma (R_2)^2 = 34.5^2 = 1190.25.$

$\Sigma R_3 = 14 + 10.5 + 15 + 12.5 + 12.5 = 64.5.$

$\Sigma (R_3)^2 = 64.5^2 = 4160.25.$

 Step 3: Calculate H obtained using the following formula:

$$H_{obt} = \left[\frac{12}{N(N+1)} \right] \left[\sum \frac{(R_i)^2}{n_i} \right] - 3(N+1).$$

Note that the capitalized N is the total N for all subjects participating in the experiment, regardless of condition, while the small n_i refers to the sample size in each group/condition, respectively. Confusing these two values is a common error, so be very careful as you work through examples on your own that you keep track of which sample size is required in the formula.

$$H_{obt} = \left[\frac{12}{15(15+1)} \right] \left[\sum \frac{(21)^2}{5} + \frac{(34.5)^2}{5} + \frac{(64.5)^2}{5} \right] - 3(15+1).$$

$H_{obt} = (0.05)(88.2 + 238.05 + 832.05) - 48.$

$H_{obt} = 57.915 - 48 = 9.915.$

 Step 4: Evaluate H obtained using the chi-square table.

You need to know the number of groups in the experiment (k) and your α level to look up the H-critical value. In this example, you have three conditions, so the $df = k - 1$ or $3 - 1 = 2$, and $\alpha = 0.05$. Under these conditions, your H-critical value is 5.99.

Decision rule: If H obtained $\geq H$ critical, then reject the null hypothesis.

Since $9.915 \geq 5.99$, we can reject the null hypothesis. The three conditions vary significantly in GPA scores, suggesting that time of day influences course performance.

Results if you use SPSS to conduct the Kruskal-Wallis test

	Class	N	Mean Rank
GPA	1.00	5	4.20
	2.00	5	6.90

(Continued)

(Continued)

	Class	N	Mean Rank
	3.00	5	12.90
	Total	15	

NPar Tests

Kruskal-Wallis Test

Ranks

	GPA
Chi-Square	9.968
df	**2**
Asymp. Sig.	**.007**

Test Statistics

a. Kruskal Wallis Test

b. Grouping Variable: CLASS

Note that SPSS reports a chi-square obtained value rather than an H-obtained value. In addition, the value is adjusted because we have a small sample size. However, SPSS uses the same degrees of freedom (based on chi-square test) that we used and calculates an exact probability (0.007) that is statistically significant, and thus the result is that same—we must reject the null hypothesis.

How would these results be reported in a scientific journal article?

If the journal required APA format, the results would be reported in a format something like this:

The three conditions vary significantly in GPA scores, suggesting that time of day influences course performance so that students in later classes perform more successfully, H ($N = 15$) $= 9.92, p < .01$.

This formal sentence includes the dependent variable (GPA), the independent variable (time of day in three categories), as well as a statement about statistical significance, the symbol of the test (*H*), the sample size (15), the statistical value (9.92), and the estimated probability of obtaining this result simply due to chance (< .01).

Assumptions for the Kruskal-Wallis (*H*) Test

1. The dependent variable (data) are measured on an ordinal or better scale.

2. The scores come from the same underlying distribution.

3. You must have at least five scores per group to use the chi-square table.

4. Random sampling

SPEARMAN'S RANK ORDER CORRELATION (rho, OR r_s)

The Spearman rank correlation test is appropriate when one or both of the variables of interest are on an ordinal scale. This test replaces Pearson's correlation test when the data are not collected on an interval or ratio scale but are at least ordinal.

Calculation Steps

1. Independently rank each variable *X* and *Y* (strongest response = 1).

2. Give tied ranks an average ranking score, but then don't reuse the ranks that you averaged. This is the same technique that we used for the Mann-Whitney *U* test and Wilcoxon when we had tied ranks.

3. Calculate the "difference in rank" scores [$D_i = R(X_i) - R(Y_i)$] and the squared differences in rank.

4. Calculate rho (r_s) using the following formula:

$$r_s = 1 - \frac{6(\sum D_i^2)}{N^3 - N}.$$

5. Evaluate r_s using Table I in the Appendix.

Complete Example

Determining the correlation between high intelligence (intelligence quotient [IQ] scores on an interval scale) and a high level of aggressiveness of an individual (measured on an ordinal scale from 1–4 with 1 being the most aggressive) can be analyzed with the Spearman rank order correlation, or rho. It is necessary to use the Spearman rank order correlation, or rho, rather than the Pearson r correlation because one of our variables is measured on an ordinal scale.

X (Aggressive Personality Scale)	Y (IQ Score)
	92
3	128
4	134
2	98
1	87
1	101
3	100
2	108

Data

 Step 1: Rank the two variables separately and handle tied ranks appropriately.

X_i (Aggression)	$R(X_i)$	Y_i (IQ Score)	$R(Y_i)$
4	7.5	92	7
3	5.5	128	2
4	7.5	134	1
2	3.5	98	6
1	1.5	87	8
1	1.5	101	4
3	5.5	100	5
2	3.5	108	3

Note that the lower scores on the aggressiveness scale mean that the individual is more aggressive, so that person gets top ranks (one and two, etc.). Similarly, the higher IQ scores imply greater intelligence, so a high IQ score receives the top rank (rank = 1 for an IQ of 134). If you were to mix these, then you'd be trying to correlate low intelligence with high aggressiveness or high intelligence with low aggressiveness. In this case, we are specifically trying to evaluate any potential correlation between high IQ and high aggressiveness, so high values for either variable should receive top ranks.

Step 2: Calculate "difference in rank" scores $[D_i = R(X_i) - R(Y_i)]$ and square the difference scores as a intermediate step in preparation for using the r_s formula.

X₁ (Aggressive)	R(X₁)	X₂ (IQ Score)	R(X₂)	Dᵢ	Dᵢ²
4	7.5	92	7	0.5	0.25
3	5.5	128	2	3.5	12.25
4	7.5	134	1	6.5	42.25
2	3.5	98	6	−2.5	6.25
1	1.5	87	8	−6.5	42.25
1	1.5	101	4	−2.5	6.25
3	5.5	100	5	0.5	0.25
2	3.5	108	3	0.5	0.25
				$\Sigma = 0$	$\Sigma = 110$

 Step 3: Calculate rho (r_s) using the following formula:

$$r_s = \frac{6(\sum D_i^2)}{N^3 - N}.$$

$$r_s = 1 - \frac{6(110)}{8^3 - 8} = 1 - \frac{660}{504} = -0.3095 \quad \text{or}$$

−0.31 (rounded to two decimal places).

Step 4: Interpret the correlation coefficient.

The Spearman rank order coefficient (r_s) for our IQ-aggressiveness example data is −0.31. A correlation coefficient this size suggests a fairly weak negative relationship between our two variables. But what we really need to do is to determine the probability that we could get a statistic of −0.31 or more extreme by chance, the same question that we have been asking throughout much of Parts II and III. We also need to keep in mind that the negative correlation suggests that top-ranking IQ scores are associated with bottom-ranking aggressiveness scores. This effect is in the opposite direction of the researcher's original hypothesis. One could argue that these researchers should have used a two-tailed hypothesis and analysis. Evaluating all hypotheses as two-tailed comparisons protects researchers from instances where the real effect is in the opposite direction that they predicted.

For Pearson's r, we were determining the probability that, given a sample coefficient of −0.31, the correlation coefficient of the underlying population is really zero, reflecting no relationship between X and Y. We evaluated the Pearson r value using a table for critical values of r based on the degrees of freedom ($N - 2$) and the desired level of alpha (see the Appendix). We will evaluate the Spearman rank order coefficient in a similar way (see Table I). In our case, the df are equal to 6 ($8 - 2 = 6$) and $\alpha = 0.05$. The two-tailed critical value is 0.886. The absolute value of the r_s obtained (0.31) calculated in this example is less than the two-tailed critical value (0.886), and the critical value is positive and your obtained value is in the opposite tail of the distribution.

Another way to express the relationship between the independent and the dependent variables is to square your r-obtained value to make it R^2.

$$r_s = \sqrt{\text{proportion of total variance of } Y \text{ that is explained by } X}.$$

$$R^2 = \text{proportion of the variability in } Y \text{ that is explained by } X.$$

The use of the capitalized R is intentional as R^2 is the formal symbol for this statistic. If we apply this concept to our example, then R^2 is the proportion of the variance in intelligence that is explained by how aggressive a personality the individual exhibits. Obviously, other possible (and potentially more likely!) factors are related to intelligence. If $r_s = -0.3095$, then $R^2 = (-0.3095)^2 = 0.09579$. This means that approximately 9.579% of the variability in intelligence is explained by aggressiveness in the individual's personality. If this were the case, then only 9.579% of the variability in your dependent variable would be explained by your independent variable, with 90.421% of the variability in your dependent variable (Y) left unexplained. This would suggest that the factor you are studying does not have a strong relationship with your dependent variable.

Results of the r_s correlation if you use SPSS:

			AGGRESS	IQ
Spearman's rho	AGGRESS	Correlation Coefficient	1.000	.342
		Sig. (2-tailed)		.408
		N	8	8
	IQ	Correlation Coefficient	.342	1.000
		Sig. (2-tailed)	.408	
		N	8	8

Correlations

Note that SPSS has calculated a slightly different value for Spearman's rho (r_s) at 0.342. This is most likely due to the way that SPSS handles tied ranks, but the value is similar to our calculated value (although SPSS does not rank scores in a manner that matches your hypothesis, so you need to be clear on the relationship between the two variables). Clearly, both our manual analysis and the computerized analysis agree that the results are not significant (SPSS assigns the significance or *p* value at .408, which is greater than .05).

How would these results be reported in a scientific journal article?

If the journal required APA format, the results would be reported in a format something like this:

There was no significant relationship between high IQ and a high level of aggressiveness in an individual, $r_s(6) = -.31, p > .05; R^2 = .10$.

This formal sentence includes the two variables (IQ score and aggressiveness score) as well as a statement about statistical significance, the symbol of the test (r_s), the degrees of freedom (6), the statistical value (−.31), and the estimated probability of obtaining this result simply due to chance (> .05).

EFFECT SIZES AND POWER

Effect size and power can be calculated for nonparametric tests. If you do not know your effect size, you can still proceed by using software and the values from your statistical output. The G*Power program that we have recommended previously can be used to calculate power for chi-square and contingency tests (www.psycho.uni-duesseldorf.de/aap/projects/gpower/).

Other Resources for Calculating Power

If you have an experimental design that does not fit with the options for calculating power with G*Power, you can use the following website to find more free resources on the Internet for calculating statistics, including power: http://statpages.org/.

This flowchart includes the new tests that were introduced in this chapter and summarizes the inferential tests that you learned in this textbook.

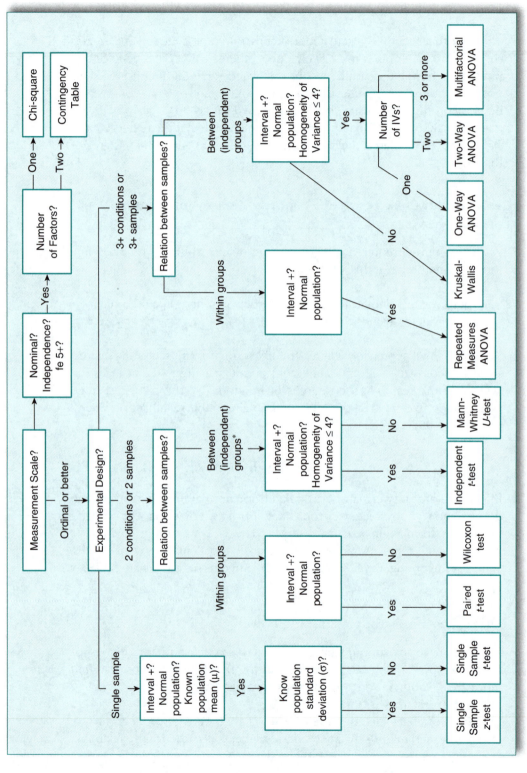

Final Flowchart on Choosing the Appropriate Test

We have placed notes regarding the major assumptions of the test in the boxes as a reminder, but you must know the rules for meeting each assumption to use this flowchart effectively. We will review the assumptions for each test in the final chapter.

SUMMARY

In this chapter, you have covered statistical tests that are appropriate for use when you do not meet the assumptions for parametric tests. We have shown you the nonparametric equivalents of the paired/correlated t test, the independent t test, and the one-way ANOVA. In addition, we discussed the chi-square and contingency table analyses for use with nominal data and the Spearman rank correlation coefficient that is appropriate when one or more of the variables are ordinal. In the next chapter, we will review the tests you have learned in this book.

CHAPTER 14 HOMEWORK

Provide a short answer for the following questions.

1. If there is a large difference between f_o and f_e for each cell, are you likely to reject or fail to reject the null hypothesis?

2. What are the assumptions for the χ^2 (chi-square) test?

3. What are the assumptions for the Wilcoxon test?

4. What are the assumptions for the Mann-Whitney U test?

5. What are the assumptions for the Kruskal-Wallis test?

6. When is it appropriate to use Spearman's rank test?

7. What do you do if f_e is less than 5?

8. What is a contingency table?

9. List three parametric tests and their nonparametric equivalent tests.

10. Chi-square is used to test differences between _____.

    A. frequencies

    B. means

    C. variances

11. A survey was conducted to determine whether there was a relationship between age (25 and younger or older than 25) and preference for Coca-Cola and Pepsi. The following data were recorded. Using an alpha level = .05, what do you conclude?

Soda Preference			
	Pepsi	Coca-Cola	Total
≤ 25 years	82	118	200
> 25 years	68	132	200
Total	150	250	400

12. A manufacturer of watches is examining preferences for digital versus analog watches. A sample of 200 people is selected and classified for age and preference. Is there a relationship between gender and watch preference? Use $\alpha = 0.05$.

	Digital	**Analog**	**Undecided**	**Total**
Male	90	40	10	
Female	10	40	10	
Total				

13. A random sample of 50 severely ill patients from a medical center was asked whether they were satisfied with their doctors. Is there a significant difference between the observed data and the expected data? Use $\alpha = 0.01$.

	Satisfied
Very	15
Not at all	35

14. Ten years ago, the Generic School District conducted district-wide reading assessments in Grades 3, 6, and 9. The results indicated that 62% of third graders, 56% of sixth graders, and 42% of ninth graders performed satisfactorily on the assessment. The assessment was repeated a decade later after new reading programs were established and 400 students in each grade level took the assessment. Has the performance changed in the past 10 years?

	Third	Sixth	Ninth
# with satisfactory scores	298	305	217

Choose the most appropriate and powerful test:

15. A developmental psychologist is examining problem-solving ability for grade school children. Random samples of 5-year-old, 6-year-old, and 7-year-old children are obtained with three children in each sample. Each child is given a standardized problem-solving task, and the psychologist records the numbers of errors.

16. Five pairs of teenagers are randomly selected and matched on a measure of general fitness. Each member of each pair is then randomly assigned to either the walking treatment or the dance treatment. After a specified time, students in each group are given a test of cardiovascular fitness. Heartbeats per minute are obtained for each subject and analyzed.

17. A researcher would like to measure the effects of air pollution on life expectancy. Two samples of newborn rats are randomly selected from a normally distributed population. The first sample of 10 rats is housed in cages where the atmosphere is equivalent to the air in a polluted city and lives for an average of

474 days ($s = 32$). The second sample of 20 rats is placed in clean air cages and lives for an average of 556 days ($s = 4$). Does pollution cause a difference in life expectancy?

18. A researcher examines the effect of relaxation training on test anxiety. Baseline levels of test anxiety are recorded for a sample of 31 subjects. All the subjects then receive relaxation training for 3 weeks and their anxiety levels are measured again. Is there a treatment effect?

Answer Questions 19 to 23 based on the story problem and table below.

A researcher is interested in whether there is a relationship between pubertal timing and gender. Participants are classified in one of the following pubertal timing classifications: (1) early puberty, (2) late puberty, and (3) on-time puberty. A random sample of 191 males and 177 females is selected. Each participant is asked into which of the three pubertal timing classifications he or she falls. Set $\alpha = 0.05$. The results are shown below in the 3×2 contingency table:

	Early Puberty	Late Puberty	On-Time Puberty	Row Marginals
Male	#1 24	#2 39	#3 128	191
Female	#4 43	#5 21	#6 113	177
Column marginals	67	60	241	368

19. State the null and alternative hypothesis.

20. State the critical value of the test statistic.

21. Find expected frequencies for each cell.

22. Calculate the chi-square.

23. State your statistical decision and draw an appropriate conclusion about the research question.

The following homework questions should be answered using the online data sets provided on the textbook website. These are meant to be conducted in a software package such as SPSS.

24. Analyze the data set using a Wilcoxon test, and indicate if you have a statistically significant result. Explain how you evaluated the output.

25. Analyze the data set using a Mann-Whitney test, and indicate if you have a statistically significant result. Explain how you evaluated the output.

26. Analyze the data set using a Kruskal-Wallis test, and indicate if you have a statistically significant result. Explain how you evaluated the output.

CHAPTER 15

Review

I n this chapter, we review the concepts and statistical tests from the previous chapters. We address how to choose the appropriate statistical analysis for the independent and dependent variables in a particular study. This is critical as most software programs do not do this for you, and your ability to evaluate the statistical analyses of others is dependent on your skills to determine which analysis matches a given experimental design. Last, we present a real-world example of a data set and demonstrate how to conduct the appropriate analyses and interpret the results using SPSS.

We have deliberately chosen a data set that is conducive to more than one analysis, which is typically the case with data sets that are from large studies and are worthy of publication. In other words, most real-world data sets are designed so that the researchers can attempt to address more than one hypothesis or question. This just means that you have more data to play with and that you will need to engage your skills in "choosing the appropriate test" for each aspect of a data set.

CHOOSE-THE-APPROPRIATE-TEST REVIEW

So far, you have primarily used the flowchart to determine the appropriate test for an analysis. Some of our undergraduate statistics students have also found an outline or table format a useful way to organize the information in the flowchart. In Table 15.1, we have organized the statistical tests that were covered in this book based on the experimental design or measurement scale (our outline format). If you include the post hoc test (Tukey), you have learned 10 inferential tests in addition to correlation, regression, and descriptive statistics.

Table 15.2 presents the choose-the-appropriate test criteria in a true "table format" with the typical assumptions of the parametric tests numbered from 1 to 5 (see also Figures 15.1 and 15.2). Note that some assumptions for the chi-square and the assumption of random sampling are not specifically addressed in this brief decision table format.

Table 15.1

Single-Sample Designs
    z test for single samples
    t test for single samples
Two Groups/Paired or Correlated
    Paired t test
    Wilcoxon signed ranks test
Two Groups/Independent
    Independent t test
    Mann-Whitney U test
Multigroup/Independent and Single Factor
    One-way analysis of variance (ANOVA)
        A priori or planned comparisons
            Independent t test
        A posteriori or post hoc comparisons
            Tukey
    Kruskal-Wallis H test
Nominal data
    Chi-square

Outline Format for Choosing the Appropriate Inferential Test

Table 15.2

# of Groups or Conditions	Type of Design	Assumptions (see numbered text)	Type of Test to Use
1 sample	Single sample	1, 2, 4, and 5 are all met	Single-sample z test
1 sample	Single sample	1, 2, and 4 are all met	Single-sample t test
2	Independent (**between**) groups	1, 2, and 3 are met	Independent t test
2	Independent (**between**) groups	At least 1, 2, or 3 broken	Mann-Whitney U
2	Dependent (**within**) groups	1 and 2 are both met	Paired t test (correlated t test)
2	Dependent (**within**) groups	At least 1 or 2 broken	Wilcoxon
1 factor		Using nominal data	Chi-square
2 or more factors		Using nominal data	Contingency table
3 or more	Independent (**between**) groups	1, 2, and 3 are met	ANOVA
3 or more	Independent (**between**) groups	At least 1, 2, or 3 broken	Kruskal-Wallis

Table Format for Choosing the Appropriate Inferential Test

Have these assumptions been met?

1. The data must be interval or ratio.

2. The data are normally distributed, meaning (a) the population raw scores are known to be normally distributed, (b) the sample size is > 30, or (c) the skewness and kurtosis values are approximately between −1.0 and +1.0.

3. The variances are equal between the groups, called homogeneity of variance (HOV). The variances can be up to four times different from each other, but no more than that, and still be considered "equal." To find HOV, you divide the larger variance by the smaller variance.

4. Known population mean

5. Known population standard deviation

Figure 15.1

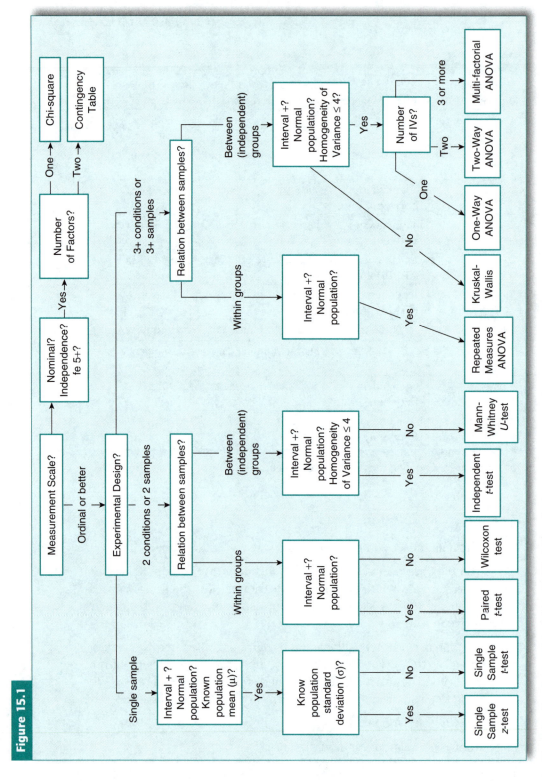

Choosing the Appropriate Inferential Test Flowchart

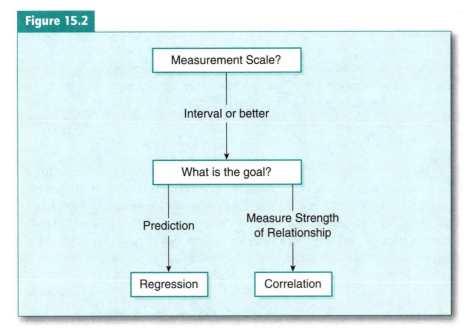

Figure 15.2

Flowchart for Choosing the Appropriate Inference Test When Independent Variable Is Continuous

TYING-IT-ALL-TOGETHER REVIEW

Real-world research often involves more than one hypothesis, independent variable, and dependent variable. Funding agencies, scientists, students, and the general public typically want to maximize the amount of information that can be obtained from a single study. This makes sense from both an academic as well as a financial perspective. For pedagogical reasons, we have chosen to minimize the complexity of many of the experimental designs and data sets that were presented in this book. Now that you are more familiar with statistics, you are ready to begin familiarizing yourself with more complex designs and analyses. Here we present a research design and data that are appropriate for more than one analysis, along with the SPSS output from those analyses.

Real-World Example

Research Questions and Design

An organizational psychologist examined the factors related to burnout or stress in engineers. Data were gathered from volunteers ($n = 50$) chosen by randomized selection from a professional technical union. The first hypothesis was that stress would be positively related to age. Exact age was determined by birth date. The second hypothesis was that gender and marital status (married or unmarried) would be related to the emotional exhaustion level of burnout/stress. Stress was measured as the level of cortisol (a hormone associated with stress) in the bloodsteam (pic/ml) and as the number of days of work missed due to illness.

Determining the Appropriate Analyses

Based on the story, Hypothesis 1 is completely separate from Hypothesis 2. Hypothesis 1 could be analyzed as a correlation because both variables (age and stress) have been measured on a continuous scale. In fact, we can perform two correlational analyses as there are two dependent variables for stress (cortisol and days sick) but only one independent variable (age). Hypothesis 2 is more complex, with two independent variables that are categorical (gender and marital status) in addition to the two dependent variables (cortisol and number of days sick). In this case, it is appropriate to use the general linear modeling function in SPSS or a similar statistical package.

SPSS Output and Interpretation for Hypothesis 1

Correlations

		AGE	CORTISOL
AGE	Pearson Correlation	1.000	.391**
	Sig. (2-tailed)	.	.005
	N	50	50
CORTISOL	Pearson Correlation	.391**	1.000
	Sig. (2-tailed)	.005	.
	N	50	50

**. Correlation is significant at the 0.01 level (2-tailed).

Exploring the Relationship Between Cortisol and Age in Engineers

It appears that age (in years) and cortisol (pic/ml) in the bloodstream are positively correlated. Note that the Pearson product moment correlation value is 0.391 (a positive number) and that the two-tailed significance of this value is 0.005. This suggests a moderate correlation between age and cortisol.

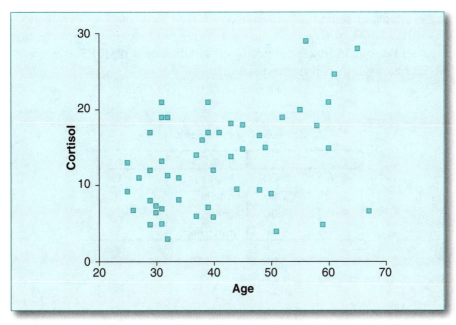

Graph

The simple scatterplot suggests a pattern that cortisol levels tend to increase with age. Of course, we must be careful to remember that correlation does not imply causation and that we cannot generalize beyond our population of interest (engineers in unions).

Correlations

		AGE	DAYSSICK
AGE	Pearson Correlation	1.000	.144
	Sig. (2-tailed)	.	.317
	N	50	50
DAYSSICK	Pearson Correlation	.144	1.000
	Sig. (2-tailed)	.317	.
	N	50	50

Exploring the Relationship Between Sick Days and Age in Engineers

It appears that age (in years) and sick days (annually) are not related. Note that the Pearson product moment correlation value is 0.144, a positive number but one in which the two-tailed significance is 0.317. This suggests no correlation between age and sick days.

Between-Subjects Factors

		Value Label	N
GENDER	1.00	Male	34
	2.00	Female	16
MARITAL	1.00	Married	29
	2.00	Unmarried	21

SPSS Output and Interpretation for Hypothesis 2

General Linear Model

SPSS automatically prints out your sample sizes and reminds you how you coded your data (1 = male and 2 = female, etc.).

Multivariate Tests [b]

Effect		Value	F	Hypothesis df	Error df	Sig.
Intercept	Pillai's Trace	.838	116.585[a]	2.000	45.000	.000
	Wilks' Lambda	.162	116.585[a]	2.000	45.000	.000
	Hotelling's Trace	5.182	116.585[a]	2.000	45.000	.000
	Roy's Largest Root	5.182	116.585[a]	2.000	45.000	.000
GENDER	Pillai's Trace	.021	.476[a]	2.000	45.000	.624
	Wilks' Lambda	.979	.476[a]	2.000	45.000	.624
	Hotelling's Trace	.021	.476[a]	2.000	45.000	.624
	Roy's Largest Root	.021	.476[a]	2.000	45.000	.624
MARITAL	Pillai's Trace	.044	1.024[a]	2.000	45.000	.368
	Wilks' Lambda	.956	1.024[a]	2.000	45.000	.368
	Hotelling's Trace	.045	1.024[a]	2.000	45.000	.368
	Roy's Largest Root	.045	1.024[a]	2.000	45.000	.368
GENDER * MARITAL	Pillai's Trace	.229	6.689[a]	2.000	45.000	.003
	Wilks' Lambda	.771	6.689[a]	2.000	45.000	.003
	Hotelling's Trace	.297	6.689[a]	2.000	45.000	.003
	Roy's Largest Root	.297	6.689[a]	2.000	45.000	.003

a. Exact statistic

b. Design: Intercept+GENDER+MARITAL+GENDER * MARITAL

The notes that SPSS has placed under the Multivariate Tests box include "b. Design: Intercept + Gender + Marital + Gender*Marital." This is the same format for the general linear model that we introduced in the GLM chapter. Both gender and marital are main effects, while Gender*Marital is the interaction. This table provides a multivariate test (overall test) of both of the dependent variables simultaneously and suggests that there are no significant main effects of gender or marital status, but there is a significant interaction of gender and marital status on overall stress (measured by cortisol and days sick). SPSS has conducted the analysis using four different multivariate techniques (Pillai's trace, Wilks's lambda, Hotelling's trace, and Roy's largest root). The subtle differences between these multivariate tests would be a topic for a more advanced course.

Tests of Between-Subjects Effects

Source	Dependent Variable	Type III Sum of Squares	df	Mean Square	F	Sig.
Corrected Model	CORTISOL	502.260[a]	3	167.420	5.354	.003
	DAYSSICK	125.081[b]	3	41.694	2.028	.123
Intercept	CORTISOL	7029.801	1	7029.801	224.792	.000
	DAYSSICK	380.436	1	380.436	18.504	.000
GENDER	CORTISOL	30.285	1	30.285	.968	.330
	DAYSSICK	.261	1	.261	.013	.911
MARITAL	CORTISOL	65.250	1	65.250	2.087	.155
	DAYSSICK	6.707E-03	1	6.707E-03	.000	.986
GENDER * MARITAL	CORTISOL	256.424	1	256.424	8.200	.006
	DAYSSICK	101.236	1	101.236	4.924	.031
Error	CORTISOL	1438.535	46	31.272		
	DAYSSICK	945.739	46	20.560		
Total	CORTISOL	10559.589	50			
	DAYSSICK	1539.000	50			
Corrected Total	CORTISOL	1940.794	49			
	DAYSSICK	1070.820	49			

a. R Squared = .259 (Adjusted R Squared = .210)

b. R Squared = .117 (Adjusted R Squared = .059)

SPSS also includes a Between-Subjects Effects table that shows the effects of each dependent variable separately from the other. Again, you should note that the main effects are not significant, but there is a significant interaction of gender and marital status on cortisol (significance or $p = .006$) and a significant interaction of gender and marital status on the number of days you call in sick (significance or $p = .031$). Interpreting the interaction would be facilitated by graphing the results and/or evaluating the descriptive statistics. Look over the means and standard deviations below to see if that helps you understand why there is no significant difference due to gender or marital status, but there is an interaction.

Descriptive Statistics

	GENDER	MARITAL	Mean	Std. Deviation	N
CORTISOL	Male	Married	10.1353	6.4893	19
		Unmarried	17.6187	4.6383	15
		Total	13.4368	6.8060	34
	Female	Married	13.4000	5.9666	10
		Unmarried	10.9350	3.4335	6
		Total	12.4756	5.1777	16
	Total	Married	11.2610	6.4038	29
		Unmarried	15.7090	5.2516	21
		Total	13.1292	6.2935	50
DAYSSICK	Male	Married	4.6842	5.7644	19
		Unmarried	1.5333	2.4456	15
		Total	3.2941	4.8149	34
	Female	Married	1.4000	2.2211	10
		Unmarried	4.5000	6.6257	6
		Total	2.5625	4.4717	16
	Total	Married	3.5517	5.0468	29
		Unmarried	2.3810	4.1289	21
		Total	3.0600	4.6748	50

While the pattern is fairly clear from the descriptive statistics, note how the graphs make the picture even clearer. Recall that crossing lines suggest an interaction!

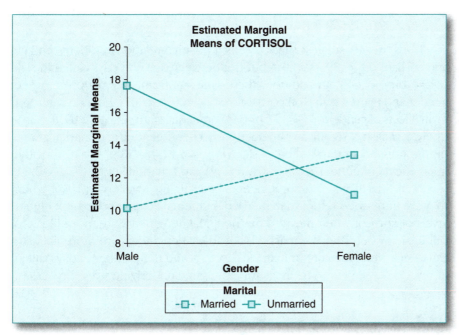

Cortisol

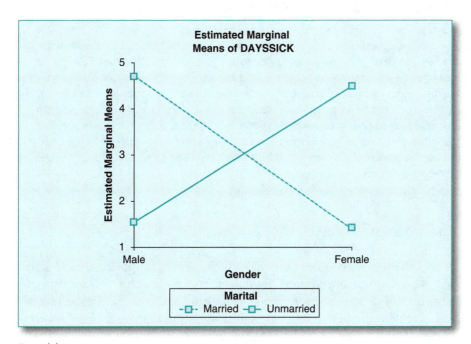

Dayssick

Conclusions From the Sample Data

These results suggest that there is a positive relationship between age and cortisol ($r_{50} = .39$, $p < .01$) but not between age and days called in sick ($r_{50} = .14$, $p = .32$). In addition, there is a significant interaction between gender and marital status on cortisol ($F_{1, 46} = 8.20$, $p < .01$), but no significant main effects of gender ($F_{1, 46} = .97$, $p = .33$) or marital status ($F_{1, 46} = 2.09$, $p = .15$). There was also a significant interaction between gender and marital status on the number of days called in sick ($F_{1, 46} = 4.92$, $p = .03$), but not for the main effects of gender ($F_{1, 46} = .01$, $p = .91$) and marital status ($F_{1, 46} = 0.00$, $p = .99$). Specifically, male engineers show higher levels of cortisol when they are unmarried rather than married, but female engineers exhibit higher levels of cortisol when they are married. Male engineers are more likely to call in sick when they are married, but female engineers are more likely to call in sick when they are unmarried. Cortisol and the number of days called in sick do not appear to be measuring the same underlying stress processes.

FYI: NOT MEETING THE ASSUMPTIONS— RESAMPLING STATISTICS

There is another option for analyzing your data when you do not meet the assumptions for parametric tests, and we have not addressed that in this text because it is a topic for advanced readers. These are techniques where you create your own sampling distribution from your raw data. This allows you to proceed even when the parametric assumptions are not met, such as when your sample distribution is not normally distributed or your variances are not homogeneous. Because these resampling statistics rely on drawing from your own data set to determine probability and variance, they allow you to avoid the otherwise important assumptions of parametric tests. Popular resampling techniques include bootstrap analysis, jackknife analysis, and permutation tests. There are a number of good references for these tests if you would like to learn more about them (e.g., Adams & Anthony, 1996; Good, 2006, 2009). They are becoming increasingly common in many disciplines.

SUMMARY

In this chapter, we reviewed the major terms and concepts from each of the following chapters; presented the flowcharts for choosing the appropriate test, including some new formats for that information; and gave you an example of a complex research design and analysis that more closely resembles statistical analysis in the real world. In our own undergraduate statistical courses, we have found that students learn these concepts at very different rates. For some students, the light bulb goes on immediately, but for others, it is near the end of the course when the concepts begin to integrate and make sense. For still others, it is later, in advanced courses, when they truly understand the basic concepts and can internalize them fully. Regardless of where you fall in that continuum, you may find that you are able to return to the concepts presented in this book later with new understanding and insights. A colleague of ours once said that learning statistics is like walking up a spiral staircase. As you circle around and gain elevation, you find that you are looking at the same view but with a new perspective. Statistics is one of those subjects where the more you learn, the more you understand the concepts that came before. As such, you may find this book is useful to you after you have completed your course in statistics. In fact, the books from our statistics courses (undergraduate and graduate) are some of the few textbooks we have kept over the years.

CHAPTER 15 HOMEWORK

Choose the most appropriate and powerful statistical test. Be sure to indicate any assumptions you may be making.

1. A researcher would like to measure the effects of a gene therapy on the life expectancy of roundworms. Two samples of roundworms are randomly selected from a normally distributed population. The first sample of 10 worms is given no gene therapy and lives for an average of 24 days ($s = 8$). The second sample of 20 worms is given a gene therapy and lives for an average of 33 days ($s = 3$). Does the gene therapy extend the life of the worms?

2. A physician is testing the effectiveness of a new arthritis drug by measuring the grip strength of 10 patients before and after they receive the drug. Grip strength is measured on a 10-point scale, and the score for each patient is calculated as the difference in strength between the two conditions.

3. A psychologist interested in the development of manual dexterity prepared a block manipulation task for 3-year-old children. A sample of 83 children was obtained, with 40 girls and 43 boys. The psychologist recorded the amount of time (sec) required by each child to arrange the blocks in a specified pattern. The girls took an average of 48 seconds, and the boys took an average of 39 seconds. The variance for the girls was 18, and the variance for the boys was 6.

4. Four wines are rated according to taste preference. Two hundred subjects are randomly selected to rate the wines. Each subject rates only one wine (50 subjects are randomly assigned to taste each wine). The rating scale is from 1 to 20, with 20 representing the highest possible score. Is there a significantly higher preference for any of the wines?

5. A dog trainer is interested in whether his puppy classes have an effect on the time it takes for his dogs to complete an obstacle course later in life. He randomly selects a sample of 60 puppies to join his class. When they are 1 year old, he records how long it takes each of them to complete a standardized obstacle course for the first time. He finds that it takes an average of 9.2 minutes ($s = 2.3$ min). He knows from extensive historical records that the average 1-year-old dog takes 10.9 minutes ($s = 3.2$ min) to complete the standardized obstacle course for the first time. Do his puppy classes have an effect?

6. A researcher is interested in whether there is an effect of alcohol on bowling ability. She records the scores of 65 subjects both before and after they consume a specified amount of alcohol. The average score before the alcohol is 115 ($s = 15$), and the average score after the alcohol is 113 ($s = 29$).

7. A researcher suspects that color blindness is inherited by a sex-linked gene. This possibility is examined by looking for a relationship between gender and color vision. A sample of 1,000 people is tested for color blindness, and then they are classified according to their sex and color vision status (normal, red-green blind, other-color blindness). Is color blindness related to gender?

	Normal	Red-Green Blind	Other-Color Blind
Male	320	70	10
Female	580	10	10

8. An investigator conducts an experiment involving the effects of three levels of a drug on memory. Subjects are randomly assigned to one of three conditions. A different drug level is administered in each condition. Memory is measured 10 minutes after each subject receives the drug, and the variances are homogeneous.

   Group means:

   A. (5-mg drug) = 1, $n = 45$

   B. (20-mg drug) = 4, $n = 45$

   C. (50-mg drug) = 10, $n = 48$

9. Researchers compared individuals who had and had not taken college-level mathematics on their knowledge of high school algebra. The mean of students on the exam who did not take math in college was 71.5, while the mean on the exam for students who did take math in college was 79.6. The variances were homogeneous. What can we conclude from these results?

10. A sports psychologist wanted to find out if pro basketball players' free throw percentage in games would improve if they shot 31 additional free throws before each game. Players were matched individually with other players on game percentage. Sixty-six players participated in the experiment. Half of the players received the treatment effect (shooting 31 additional free throws before each game), and the other half served as controls. Researchers then compared game free throw percentage between the conditions.

11. Researchers wanted to see if there is a relationship between faculty and students on how much they like the tag line **LEARNING AT THE LEADING EDGE**. Researchers found that 50 students liked the slogan, 15 did not, and 35 had no preference. Forty-five faculty members liked the slogan, 15 did not, and 40 had no preference. The variances were homogeneous. How would you analyze the data?

12. A survey was conducted to find out how well sociology graduate students liked learning sociology at the University of Washington compared to where they received their undergraduate degrees. Twenty-eight sociology graduate students were surveyed. Twelve preferred their undergraduate school, and 16 preferred the University of Washington. The data were mutually exclusive.

13. Researchers analyzed how long bartenders ($n = 200$), sales people ($n = 350$), and clerical workers ($n = 150$) could hold their breath under water. Individuals were randomly selected from a normal population. The variances were .37, .26, and .12.

14. Researchers hypothesized that after an observer rat interacted with a demonstrator rat that had a lemon extract on its fur, the observer rat would prefer lemon over vanilla extract. To conduct their study, they gave 39 observer rats an initial preference test between vanilla and lemon where they measured the time spent on each scented side of a testing apparatus. They allowed all of the observer rats to interact with a demonstrator rat smelling of lemon. Then they conducted a second preference test where they recorded the amount of time rats spent on the lemon side after the social interaction compared to the time they spent on the lemon side in their initial preference test. How can we analyze these data?

15. Researchers wanted to know if exam scores would change if a nursing class was held in the afternoon compared to the morning. Researchers paired students in two sections, one in the morning and one in the afternoon. The section instructor, GPA, and other extraneous factors were controlled. The researcher recorded the scores in each section. How can researchers analyze the data?

Homework Answers

CHAPTER 2 HOMEWORK ANSWERS

1. A population is the entire possible data set. A sample is a subset of that data.

2. A variable is any property that may have different values at different times, under different conditions. A constant is a fixed value.

3. An independent variable is the variable being manipulated or changed; it is the cause of an effect. A dependent variable is the variable being measured; it is the effect (i.e., it is the variable that you look at to determine whether your independent variable had an effect).

4. A statistic is a summary calculation on *sample* data. A parameter is a summary calculation on *population* data.

5. A continuous variable can have an infinite number of values between each unit. A discrete variable has no values between each unit.

6. Descriptive statistics are used to summarize your sample data with a single number, such as a mean or range, while inferential statistics are numbers that we calculate based on sample data but that we are attempting to use to generalize to a larger population of data.

7. Random sampling (i.e., selecting subjects with no bias) allows us to use the laws of probability on the data we collect. It also permits us to make decisions about how representative that data are of the population.

8. There are four measurement scales that are used to collect data, and which one you use determines the mathematical manipulations that you can make on the data later.

9.

What is the independent variable?	Hours of sleep per night (4 or 8)
What is the population in this study?	38,000 University of Washington undergrads
What is the dependent variable?	Lecture attendance
What is the sample?	350 undergrads
What statistic is used?	Mean lecture attendance

10.

What is the independent variable?	Gender of the teacher
What is the population in this study?	California high schoolers
What is the dependent variable?	Test scores
What is the sample?	150 high schoolers from Santa Monica
What statistic is used?	Mean test scores

11. Nominal data are categorical. There is no numerical relationship on which to base any order or ranking.

12. Ordinal data are rank-ordered data but do not have equal intervals between data points or any measure of magnitude.

13. Interval data have equal intervals and magnitude between data points but do not have an absolute zero.

14. Ratio data are like interval data, only with an absolute zero.

15. Nominal

16. Ratio, interval

17. Ordinal: A 1–7 ranking about how helpful these questions are

18. a. 7, b. 303, c. 16,343, d. 91,809, e. 261, f. 315

19. Ratio

20. Ordinal

21. Interval

22. Interval

23. Interval

24. B. nominal

25. D. self-rating of anxiety level by students in a statistics class—ratio

CHAPTER 3 HOMEWORK ANSWERS

1. Data are collapsed into classes or ranges of values, and then the number of observations (or frequency) of each class, or range of values, is recorded in a table or graph.

2. The frequency of scores in that class interval divided by the total number of scores in the distribution.

3. Histograms are used to plot frequencies using a separate bar for each score.

4. Typically a two-dimensional graph wherein one dimension represents the independent variable and the other represents the dependent variable. The magnitude of the dependent variable is represented by the height of a rectangle with a uniform width (a bar).

5. The ordinate is the Y-axis or the axis in the vertical position.

6. A graphical display of pairs of data points (X and Y pairs) that is intended to reveal any potential relationship between the two variables.

7. A distribution that is identically shaped on either side of the central point, or mirror images of one another.

8. Relative frequency is the frequency of scores in that class interval divided by the total number of scores in the distribution, whereas cumulative frequency is the sum of the frequency of scores starting from the lowest class interval and working up to the highest class interval. The cumulative frequency (or sum of the frequencies for each class interval) for the highest class interval should be equal to the sample size (n).

9. A histogram is basically a frequency distribution with the score on the X-axis and the frequency on the Y-axis except that the frequencies are noted as rectangular bars. A bar graph differs from the histogram in that dimension represents the independent variable and the other represents the dependent variable. The magnitude of the dependent variable is represented by the height of a rectangle with a uniform width (a bar).

10. General rules of thumb include starting both axes at 0, filling the area of the graph so that it is approximately ¾ filled, and having the height at approximately ¾ of the width of the graph. If it is not feasible to have both axes begin at 0, then it is traditional to show a break on the appropriate axis with "//" to indicate that part of the scale was dropped to avoid misinterpretation of the scale. Finally, it is important to label the X-axis and the Y-axis and to give a title for the graph so that it is clear what is being represented.

11. It is best to use a scatter plot when you have pairs of data points (X and Y pairs) that are intended to reveal any potential relationship between the two variables. Histograms or frequency distribution graphs are best used when the Y-axis is the frequency of the score on the X-axis.

12. It must be positively or negatively skewed.

13. B. negatively skewed

14. A. positively skewed

15. C. kurtotic

16.

Class Intervals	Freq.	Relative Freq.	Cumulative Freq.	Cumulative %
100–109	1	0.0417	24	1.0000
90–99	3	0.1250	23	0.9583
80–89	11	0.4583	20	0.8333
70–79	7	0.2917	9	0.3750
60–69	2	0.0833	2	0.0833
Total Scores =	24			

17.

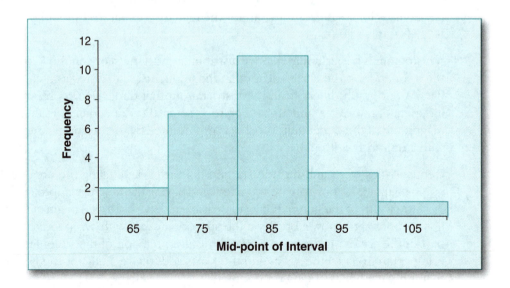

18. The distribution is nearly symmetrical but with a slight positive skew.

19.

Class Intervals	Freq.	Relative Freq.	Cumulative Freq.	Cumulative %
600–649	1	0.0200	50	1.0000
550–599	4	0.0800	49	0.9800
500–549	8	0.1600	45	0.9000
450–499	12	0.2400	37	0.7400
400–449	10	0.2000	25	0.5000
350–399	8	0.1600	15	0.3000
300–349	4	0.0800	7	0.1400
250–299	3	0.0600	3	0.0600
Total Scores =	50			

20. A. positively skewed

21. Range: $37 - 6 = 31$
~ 10 intervals: $i = 31/10 = 3.1 = 3$

Interval	f	Relative f	Cumulative f	Cumulative %
36–38	1	.03	30	100.00
33–35	2	.07	29	96.67
30–32	4	.13	27	90.00
27–29	0	.00	23	76.67
24–26	2	.07	23	76.67
21–23	4	.13	21	70.00
18–20	3	.10	17	56.67
15–17	4	.13	14	46.67
12–14	5	.17	10	33.33
9–11	3	.10	5	16.67
6–8	2	.07	2	6.67
	$N = 30$	1.0		

22.

Interval	f	Relative f	Cumulative f	Cumulative %
95–99	1	.03	30	100.00
90–94	3	.10	29	96.67
85–89	3	.10	26	86.67
80–84	3	.10	23	76.67
75–79	1	.03	20	66.67
70–74	4	.13	19	63.33
65–69	2	.07	15	50.00
60–64	3	.10	13	43.33
55–59	4	.13	10	33.33
50–54	2	.07	6	20.00
45–49	2	.07	4	13.33
40–44	2	.07	2	6.67
	$N = 30$	1.0		

CHAPTER 4 HOMEWORK ANSWERS

1. A statistic that measures the middle, or central tendency, of a set of scores and is calculated by taking the sum of the scores and dividing by the number of scores.

2. The value at which 50% of the scores fall below it.

3. The most frequent score in a set of scores.

4. A distribution that only has one mode, or most frequent score.

5. A distribution is symmetrical if it is identical on each side of the mean (mirror images).

6. While both are measures of the average dispersion of scores around the mean, the variance is in squared units while the standard deviation is not.

7. A positive skew occurs when one tail of the distribution is longer in the direction of higher scores, while a negative skew occurs when one tail of the distribution is longer toward the lower scores.

8. When you estimate the population standard deviation from a sample standard deviation, it tends to underestimate the true variability of the population. If you subtract 1 from your sample size, it provides an adjustment that ensures that your estimate of the population standard deviation is not too small. Thus, the sample standard deviation is considered a biased estimate of the population standard deviation, but the sample mean is an unbiased estimate of the population mean. If the sample mean were a biased estimate, it would require a correction, but it is an unbiased estimate of the population mean and no correction is required.

9. If you subtract every score from the mean of the collective scores and then sum those deviations, it will add up to zero. This is logical since the mean should arithmetically be the middle score.

10. We divide by the number of scores so that you end up with the *average* deviation around the mean.

11a. $\mu = 20.37$, SS = $3455 - (163^2/8) = 133.87$, $\sigma = 4.09$

11b. $\bar{X} = 20.37$, $s = 4.37$

11c. $s^2 = 19.12$

12. $\bar{X} = 26.50$, $s = 1.41$

13a. $269/12 = 22.42$

13b. Average of two centermost points $(17 + 19)/2 = 18$

13c. 19

13d. 12

13e. 22.56

14a. $643/9 = 71.44$

14b. $s = 5.66$

14c. $s^2 = 32.03$

15. Mean $= \dfrac{\sum X}{N} = \dfrac{117}{10} = \mathbf{11.7}$

16. Median $= \dfrac{11 + 12}{2} = \mathbf{11.5}$

17. Mode $= \mathbf{13}$

18. $N = \mathbf{10}$

19. $\sum X = \mathbf{117}$

20. $\left(\sum X\right)^2 = \mathbf{13,689}$

21. $\sum X^2 = \mathbf{1,483}$

22. $\text{SS} = \sum X^2 - \dfrac{\left(\sum X\right)^2}{N} = 1,483 - \dfrac{13,689}{10} = 1,483 - 1,368.9 = \mathbf{114.1}$

23. Estimate the population standard deviation =

$$s = \frac{SS}{N - 1} = \frac{114.1}{10 - 1} = \frac{114.1}{9} = 12.68 = \mathbf{3.56}$$

CHAPTER 5 HOMEWORK ANSWERS

1. A number between 0 and 1 that indicates the likelihood of an event.

2. 5/6 that the event will not happen or 1/6 that the event will happen.

3. 2/3 that the event will happen and 1/3 that the event will not happen.

4. Infinity—you would never reach that point because the curve is asymptotic to the horizontal axis.

5. Probabilities vary between 0 and 1.

6. A priori probability refers to a probability value deduced from reason alone, while a posteriori refers to an actual probability value that is calculated after collecting the data.

7. At z scores of −1 and +1 (corresponding to −1 and +1 standard deviation)

8. Z scores are data points transformed into standard scores with common units. Those common units are expressed as standard deviations from the mean.

9. −1.49, 0.55, −0.08, −0.47, 1.49

10a. $X = 45.90$

10b. $z = 1.58$, so the first subject did better because the z score is larger.

11. 8.44

12. −0.94

13. 51.31 and 56.69

14. 28.43%

15. 24.99%

16. 1.87

17. 79.07

18. The most extreme 5% of the scores fall below 98.64 and above 161.36.

19. The raw score that lies at the 90th percentile is 615.20.

CHAPTER 6 HOMEWORK ANSWERS

1. All the values that the statistic can take and the probability of getting each value under the assumption that chance alone is acting.

2. All possible values of the mean of samples of size n, with the probability of getting each value, assuming that sampling is random from the null hypothesis population.

3. Regardless of the shape of the population of raw scores, the sampling distribution of the mean approaches a normal distribution as sample size n increases.

4. Each sample mean can be considered an estimate of the mean of the raw score population. This estimate is not precise, and thus there is some sampling error associated with this estimate.

5.

   1. Determine all possible samples of size n that can be drawn from the population.

   2. Calculate the statistic that you are using for each sample.

   3. Calculate the probability of each statistical value based on chance alone. Essentially you are determining how often each statistical value occurs by chance.

6.
$$z = \frac{1.7 - 2.3}{\frac{.2}{\sqrt{12}}} = -10.40$$

$$z = \frac{15 - 34}{\frac{8.2}{\sqrt{12}}} = -8.03$$

$$z = \frac{99 - 125}{\frac{12.3}{\sqrt{12}}} = -3.55$$

7. Best: object permanence; worst: Apgar. Larger negative z scores are further below the mean.

8. a. $z(10\%) = +1.28 \ X = 2.3 - (.2)(1.28) = 2.044$.

   b. 2.0 is critical score: actual score is 1.7, so at risk.

9.
$$z = \frac{28 - 23.6}{\frac{2.30}{\sqrt{6}}} = 4.69$$

Table A, column C $= p < .0001$

10. Yes

11.

$$\sigma\bar{X} = \frac{9.00}{\sqrt{100.0}} = \frac{9.00}{10.00} = \mathbf{0.9000}$$

$$z = \frac{90.00 - 88.00}{0.9000} = \mathbf{2.22}$$

Table A, column B = .4868 + .5000 = .9868, so p = .9868.

12.

$$\sigma\bar{X} = \frac{15.00}{\sqrt{5.00}} = \frac{15.00}{2.2361} = \mathbf{6.7081}$$

$$z = \frac{163.00 - 133.00}{6.7081} = \mathbf{4.47}$$

Table A, column C < .0001, so p < .0001.

13.

Sample mean = 317   population mean = 310
n = 80   population std. dev. = 48

*Know population standard deviation so choose single-sample z test.

$$Z_{obt} = \frac{\textbf{sample mean} - \boldsymbol{\mu}}{\boldsymbol{\sigma}/\textbf{sqr. root of n}} = \frac{317 - 310}{48/\textbf{sqr. root of } 80} = \frac{7}{5.367} = \mathbf{1.30}$$

14. On the basis of the probability found in the table, we could conclude that it is unlikely that this reduction program decreased participants' anxiety levels.

15. No

CHAPTER 7 HOMEWORK ANSWERS

1. A real effect is the actual effect of the independent variable, which is not known (i.e., it might be masked by confounds).

2. Barring other factors (e.g., increasing N), power changes directly with alpha. For example, as alpha decreases, power decreases (because power $= 1 - \beta$, and as alpha decreases, β increases).

3. Logistically, we usually have to sample without replacement, but our statistical inferences require sampling with replacement.

4. So that the data are representative of the underlying population, and therefore we can use the laws of probability to make inferences.

5. A sample is random if each possible sample of a given size has an equal chance of being selected and all members of a population have an equal chance of being selected into the sample.

6. Increasing N increases power.

7. Some factors include sample size (N), the type of statistical test used, the magnitude of the real effect, the experimental design, and the alpha level.

8. Decreasing alpha increases the probability of a Type II error (β)—accepting the null hypothesis when it's actually false—because they are inversely related.

9. You could argue that it would be reasonable to set the alpha level higher in this situation given the real-world consequences. Minimizing Type II errors may be more important in this case than minimizing Type I errors because you really don't want to retain the null (the drug has no effect) if it's false.

10. Parametric statistical tests require assumptions about the parameters of the population while nonparametric tests do not require assumptions about the population.

11. A. power increases

12. C. power decreases

13. C. .95

14. B. minimizes beta

15. A. decrease, increase

16. D. All of the above

17. marital therapy affects happiness

18. marital therapy does not affect happiness

19. Reject H_0: marital therapy affects happiness

20. married couples in New York City

21. Type I error because it is possible that marital therapy does not affect happiness.

22. Negative verbal feedback affects test performance.

23. Negative verbal feedback has no effect on test performance.

24. Negative verbal feedback had a significant effect on test performance, $p < .05$.

25. Afternoon exam scores will be higher than morning exam scores.

26. There will be no difference between afternoon and morning exam scores, or exam scores will be higher in the morning.

27. Since the effect is in the direction opposite the original prediction, the null hypothesis is retained. Afternoon exams do not increase scores above morning exams.

CHAPTER 8 HOMEWORK ANSWERS

1. All possible values of the mean of samples of size N, with the probability of getting each value, assuming that sampling is random from the null hypothesis population.

2. It is used to generate the sampling distribution that allows us to test the validity of the null hypothesis.

3. Single-sample t test

4. Three of the following: interval or ratio data, random sampling (assumed for all hypothesis testing), normal distribution of means, single sample and known population mean and standard deviation.

5. B. if $t_{obt} > t_{crit}$

6. E. df

7. D. approaches the normal distribution

8. A. A and B

9. A. t test

10. D. the t distribution is a family of curves

11. B. $n - 1$

12. Single-sample t test, t obtained = –2.07, t critical = –2.423, *fail to reject* null. There was no significant reduction in page length (or there was an increase).

13. A. standard error = 1.0041, z obtained = 4.48, z obtained > z critical, *reject* null hypothesis. Note: Another way to come to the same solution is to try and look up a z obtained of 4.48. The chart doesn't go that high, but given as high as it does go, then $p < .0001$.

13. B. standard error = 0.6755, z = 1.0067, Table A, column C, p = .33, *fail to reject* null hypothesis.

14. $N = 45$ (normal distribution), interval or better data, single sample, known population mean, unknown population standard deviation: single-sample t test.

15. $N = 35$ (normal distribution), single-sample mean, known parameters of null hypothesis population: single-sample z test.

16.

Sample mean = 317   population mean = 310

$s = 37$

$n = 80$

*Do not know population standard deviation so choose single-sample t test.

$$T_{obt} = \frac{\text{sample mean} - \mu}{s/\text{sqr. root of n}} = \frac{317 - 310}{37/\text{sqr. root of } 80} = \frac{7}{4.1367} = 1.692$$

17. $df = n - 1 = 80 - 1 = 79$

    t_{crit} ($\alpha = 0.05$, $df = 79$, one-tailed) $= -1.671$

18. Fail to reject the null hypothesis. The anxiety reduction intervention program failed to decrease participants' anxiety levels.

19.

    95% CI
    upper bound: x bar + t crit × std. error $= 317 + 2.00 \times 4.1367 = 325.27$
    lower bound: x bar − t crit × std. error $= 317 - 2.00 \times 4.1367 = 308.73$
    95% probability that the population mean lies between 308.73 and 325.75.

    99% CI
    upper bound: x bar + t crit × std. error $= 317 + 2.66 \times 4.1367 = 325.27$
    lower bound: x bar − t crit × std. error $= 317 - 2.66 \times 4.1367 = 308.73$
    99% probability that the population mean lies between 306 and 328.

20. The 99% CI is broader than the 95% CI. This makes sense because the probability that the population mean would be in this broader range should be greater than the narrower range of the 95% CI.

21. Directional

22.

    Sample mean = 46   population mean = 43.75
    $s = 3.5$
    $n = 41$
    *Do not know population standard deviation so choose single-sample t test.

$$T_{obt} = \frac{\text{sample mean} - \mu}{s/\text{sqr. root of n}} = \frac{46 - 43.75}{3.5/\text{sqr. root of } 41} = \frac{2.25}{0.547} = 4.12$$

23. $df = n - 1 = 41 - 1 = 40$

    t_{crit} ($\alpha = 0.05$, $df = 40$, one-tailed) $= 1.684$

24. Reject the null hypothesis. On average, 5-year-olds today are taller than in previous years.

25. Single-sample z test

CHAPTER 9 HOMEWORK ANSWERS

1. These designs reduce the effects of individual differences among subjects (which you are not likely to be studying). Reducing variability serves to increase the power of the test.

2. The sampling distribution of difference scores must be normal, and the dependent variable must be on the interval or ratio scale.

3. The sampling distribution of means must be normal, the samples must be drawn from populations having equal variances, and the dependent variable must be on the interval or ratio scale.

4. You could know that the population from which the scores come is normally distributed, you could use the central limit theorem and assume that samples ≥ 30 are normally distributed, and you could check that the skewness and kurtosis values for your data are not significantly different from normal, and if they are different from normal, you could try to log transform them.

5. With the paired t test, the subjects are the same in each group or they are "matched" on the characteristics that might affect the experiment. Thus, we can assume that both samples come from the same population, and any differences we see are due to our independent variable. However, the independent t test uses random sampling to assign subjects to each group, and we need to assume that both groups came from the same null hypothesis population with the same variance prior to our study.

6. Drop the subjects for which you have missing data points or do an independent t test on all of your data points.

7. The paired t test can be replaced with a nonparametric test called the Wilcoxon signed ranks test, while the independent t test can be replaced with a nonparametric test called the Mann-Whitney U test. Another option is to perform a log transformation on your data to determine if the mathematical transformation will result in a normal distribution.

8. Independent t test where t obtained $= -2.062$, t critical $= +2.306$. So, fail to reject null; being raised in a military or nonmilitary family does not affect the number of happy childhood memories you can recall.

9. Paired t test where t obtained $= 3.13$, t critical $= +2.365$. So, reject the null hypothesis. Race affects whether or not you are likely to shoot an innocent bystander.

10. Independent *t* test where *t* obtained = 1.542, *t* critical = 2.110. So, fail to reject null; the day of the week does not appear to affect performance on exams.

11. $n_1 = 31$ and $n_2 = 35$ (normal distribution), two independent groups, interval or better data, $s_1 = 1.34$ and $s_2 = 2.15$ [$(2.15)^2/(1.34)^2 < 4$], so we have homogeneous variances: independent *t* test.

12. Two normally distributed paired samples, interval or better data: paired *t* test.

13. The antismoking campaign does not affect the smoking patterns of teenagers.

14. The antismoking campaign affects the smoking patterns of teenagers.

15. Nondirectional

16. It is a repeated-measures design because the study used the same participants for each condition. The participants served as their own controls.

17. The independent variable is smoking campaign (with two levels: before and after). The dependent variable is the count (frequency or number) of increases (or decrease) in the smoking categories.

18. Nominal because the data are classified into smoking either more or less; essentially, the classification records the frequency of smoking more or less.

19. Paired *t* test

20. Which of the following are assumptions underlying the use of the paired *t* test?

    A. The variance of the population is known

    B. The sampling distribution is normal

    C. Data are interval or ratio

    D. All of the above

    E. A and B

    F. B and C ***

    G. A and C

21. **INDEPENDENT *T* TEST**

$$t_{obt} = \frac{mean_1 - mean_2}{\sqrt{s_w^2(1/n_1 + 1/n_2)}} \quad \text{or}$$

$$t_{obt} = \frac{mean_1 - mean_2}{\sqrt{(((SS_1 + SS_2)/(n_1 + n_2 - 2)) * ((1/n_1 + 1/n_2)))}}$$

$$s_w^2 = \frac{(df_1 * s_1^2 + df_2 * s_2^2)}{df_1 + df_2} = 1.9834 \qquad SS = \sum x^2 - \frac{(\sum x)^2}{n} \qquad SS_1 = 14.1$$

$$SS_2 = 21.6$$

$$t_{obt} = \frac{2.7 - 3.8}{\sqrt{(1.9834 * (1/10 + 1/10))}} = -1.75 \qquad t_{crit} = \pm 2.101 \qquad t_{obt} < t_{crit},$$

so fail to reject H_0

CHAPTER 10 HOMEWORK ANSWERS

1. The variability among means increases relative to the variability within groups.

2. True, because F is a ratio of variance estimates (s_B^2/s_W^2), and since variances are always squared numbers, they are always positive, and the ratio of two positive numbers is always positive.

3. For unplanned or post hoc comparisons, this test maintains your alpha level by adjusting the probability (critical value) for the number of means being compared (r).

4. If there is no effect, then F obtained will equal 1.00 (approximately). F obtained will equal 1.00 when there is no treatment effect because you are dividing one estimate of population variance by a second estimate of the same population variance, and any number divided by itself is equal to 1.00. Of course, F obtained will rarely equal exactly 1.00 since our measurements of each estimate will vary due to random effects, but F-obtained values near 1.00 will never be significant because they will be suggesting little or no effect of our experimental treatment.

5. Both distributions form a family of curves based on degrees of freedom.

6. Use an independent ANOVA when you have more than two samples and a between-groups design, when the sampling distribution is normally distributed, when the dependent variable is on an interval or ratio scale, and when the variances of the groups are homogeneous.

7. Fourfold rule.

8.

Source	SumsSquares	df	(MS) s^2	F_{obt}	F_{crit}
Between	148.17	2	74.085	10.2976	4.26
Within	64.75	9	7.1944		
Total	212.92	11			

Reject null hypothesis

$Q_{obt} = 0.3728$ $Q_{obt} = 4.2875*$ $Q_{obt} = 4.6603*$

Group C differs from A and B.

9.

Source	SumsSquares	df	(MS) s^2	F_{obt}	F_{crit}
Between	3.2661	3	1.0887	2.32	2.72
Within	37.24	80	0.47		
Total	40.5061	83			

Fail to reject null hypothesis

10. One-way ANOVA because there are three independent groups, one variable, random sampling, normal distribution (all groups, N is greater than or equal to 30), interval data (pitch discrimination is assessed based on the students' ability to tell if a note is separated by seven or four half-steps), homogeneous variance.

11. Single-sample t test because there is a single-sample design in which the t assumptions of normality (16 subjects from a normally distributed population), random sampling, and interval or better data are met, but we do not know the population standard deviation.

12. Single-sample z test because we have both the population mean and standard deviation, a sample mean, and random sampling of a normally distributed population ($n = 30$).

13. Independent t test because there are two independent groups in which the independent t test assumptions of normality ($n = 35$ for both groups), interval or better data (number of problems correct), random sampling, and homogeneous variances are met.

14.

H_0: There is no difference in effectiveness between the weight loss programs.

H_1: One or more of the programs has a different effect on weight loss than one or more of the others.

15.

Source	SS	Df	s^2	F_{obt}	F_{crit}
Between	419.48	4	104.87	15.4685	2.76
Within	169.49	25	6.7796		
Total	588.97	29			

16. $F_{obt} > F_{crit}$ Therefore, reject H_0.

17. Tukey's

*Rank groups smallest to largest.

	5	1	3	2	4
$\bar{X}$	3.00	6.33	8.67	9.50	14.33
$\bar{X} - \bar{\bar{X}}$		3.33	5.67	6.50	11.33
			2.34	3.17	8.00
				0.83	5.66
					4.83
Q_{obt}		3.13	5.33	6.11	10.66
			2.20	2.98	7.53
				0.78	5.32
					4.54

$Q_{crit} = 4.17$

$\alpha = .05, df = 25, k = 5$

$df_B = k - 1 = 5 - 1 = 4 \qquad df_W = N - k = 30 - 5 = 25 \qquad df_T = df_B + df_W = 29$

$Q_{obt} = \bar{X} - \bar{X}/\sqrt{(S_W^2/n)}$

(denominator) $\sqrt{(S_W^2/n)} = \sqrt{(6.7796/6)} = 1.0630$

Conclusions: There are significant differences between Program 5 and Programs 3, 2, and 4 (not 5 and 1), as well as between Program 4 and all other programs.

18.

H_0: Sleep deprivation does not affect problem-solving ability.

H_1: One or more of the sleep deprivation groups has an effect on problem-solving ability.

19.

Source	SS	df	S^2	F_{obt}	F_{crit}
Between	16	2	8	3.79	4.26
Within	19	9	2.11		
Total	35	11			

20. $F_{obt} < F_{crit}$ Fail to reject.

21. There is no difference in problem solving between the three groups.

22. There is no need to perform a post hoc comparison.

23.

Source	SS	df	s^2	F_{obt}	F_{crit}
Between	18.72	2	9.36	6	4.26
Within	14	9	1.56	X	X
Total	32.72	11	X	X	X

$$F_{obt} = \frac{s_B^2}{s_w^2} \qquad s_B^2 = 1.56\,(6) = 9.36 \qquad s_B^2 = \frac{SS_B}{df_B} \qquad SS_B = 2\,(9.36) = 18.72$$

Table C: $F_{crit} = 4.26$

24. Should get same answers as in the table,

$$F_{obt} = \frac{s_B^2}{s_w^2} \qquad s_B^2 = 1.56\,(6) = 9.36 \qquad s_B^2 = \frac{SS_B}{df_B} \qquad SS_B = 2\,(9.36) = 18.72$$

Table C: $F_{crit} = 4.26$

CHAPTER 11 HOMEWORK ANSWERS

1. A repeated-measures ANOVA is used when you have a within-groups design, while a one-way ANOVA is used when you have a between-groups (independent) design.

2. Use a two-way ANOVA when you have more than two independent variables and a between-groups design, the sampling distribution is normally distributed, the dependent variable is on an interval or ratio scale (although some statisticians believe that this is too strict), and the variances of the groups are the same or are homogeneous.

3. Use a repeated-measures ANOVA when you have one independent variable with more than two levels and a within-groups design, the sampling distribution is normally distributed, and the dependent variable is on an interval or ratio scale (although some statisticians believe that this is too strict).

4. They are the same EXCEPT that you have TWO independent variables with the two-way ANOVA, which means that you calculate THREE *F*-obtained values (two main effects and one interaction).

5. It is an equation that is presented in the chapter to aid the conceptual understanding about the multiple factors that can influence an individual's score (including real effects of the independent variables that we measure as well as unexplained variance).

6. Because you are able to determine how much of the previously unexplained variance (or error) can be explained by individual differences. Having multiple measures on the same individual allows us to calculate this value. Reducing the unexplained variance means that we are able to reduce the size of our denominator (s_w^2), thus resulting in a larger *F*-obtained value (and increased power to detect any real effects of our experimental treatment).

7. A "main effect" is the influence of one of your independent variables on your dependent variable.

8. When the effect of one of your independent variables (or factors) is not the same at all levels of your other independent variable (or factor), then you have a statistical "interaction" effect.

9. Interaction terms occur when you have more than one independent variable. In the case of the repeated-measures ANOVA presented in this chapter, you only have one independent variable. The Subj variable is actually a subset of the old "within-groups variance" or error, and it would not be appropriate or interesting to determine interactions between "error" and your independent variable.

10.

ANOVA					
Source of Variation	SS	df	MS	F	Fcrit
Row (sex)	828.8167	1	828.8167	3.030429	4.01954
Columns (treatment)	1265.633	2	632.8167	2.313788	3.168246
Interaction (Sex × Treatment)	847.2333	2	423.6167	1.548883	3.168246
Within	14768.9	54	273.4981		
Total	17710.58	59			

You must fail to reject the null hypotheses for the row, column, and interaction values. There are no significant effects of your treatment or for sex.

11.

ANOVA					
Source of Variation	SS	df	MS	F	Fcrit
Row (sex)	495.0625	1	495.0625	1.797164	4.012975
Columns (status)	1791.625	3	597.2083	2.167971	2.769433
Interaction (Sex × Status)	1855.063	3	618.3542	2.244734	2.769433
Within	15426.25	56	275.4688		
Total	19568	63			

You must fail to reject the null hypotheses for the row, column, and interaction values. There are no significant effects of your treatment or for sex.

12.

ANOVA					
Source of Variation	SS	df	MS	F	Fcrit
Row (job)	560952.5	1	560952.5	2.242676	4.08474
Columns (blood pressure)	10028590	3	3342863	13.36469	2.838746
Interaction (Job × Blood Pressure)	199493.2	3	66497.74	0.265857	2.838746
Within	10005058	40	250126.4		
Total	20794093	47			

You must fail to reject the null hypotheses for the row and interaction values. There are no significant effects for job or for the interaction between job and blood pressure. However, there are significant differences between your blood pressure groups.

13. One-way ANOVA. One IV (school) with five levels and a between-groups design. Assumptions are met (interval or ratio test scores, sample sizes large enough to assume normality, homogeneity of variance was assumed in the instructions).

14. Independent *t* test. Only two groups of subjects in a between-groups design. The same size is large enough that normality can be assumed and the variances between the groups are homogeneous. The number of details is an interval or ratio scale of measurement.

15. Repeated-measures ANOVA. One group of 35 individuals was eventually analyzed for four time periods. So you had four scores for each person. The sample size is sufficient to assume normality. The homogeneity of variance assumption was thrown in as a distracter. It doesn't matter if the variances are homogeneous for this test. The memory scores were interval or ratio data.

16. $Score = \mu + IV + E$

17. $Score = \mu + IV_1 + IV_2 + (IV_1)(IV_2) + E$

18. $Score = \mu + IV_1 + Subj + E$

CHAPTER 12 HOMEWORK ANSWERS

1. It is a measure of how much variance in variable Y can be accounted for by variable X. It is calculated by squaring the value of the correlation between X and Y (r^2).

2. $r^2 = (0.89)^2 = 0.7921 = 0.79$

3. It tells us that 79% of the variability found in Y can be accounted for by X.

4. From −1.00 to 1.00

5. The shape of the relationship between X and Y and the measurement scale underlying the data

6. That the relationship between X and Y is linear.

7.
$$\sum X = 19$$
$$\sum X^2 = 71$$
$$\sum XY = 1,782$$
$$\sum Y = 1,035$$
$$\sum Y^2 = 108,293$$
$$r = -0.91$$

8. $Y' = b_y X + a_y; \ Y' = -5.286533(X) + 113.54441$

9. C. both a and b are true

10. B. a relationship exists, but all of the points do not fall on the line

11. A. higher the correlation

12. D. 0.15

13. E. None of the above

14. B. They should fire that sorry statistician and hire a new one (of course, an answer of "−1.08" is impossible!)

15a. −0.52

15b. −0.04

15c. 100.61

15d. 85.33

15e. The standard error of the estimate (SEE) = 8.22 and is quite large. The R^2 is approximately 0.27, so reaction time only explains 27% of the variability in vocabulary scores. I wouldn't be very confident about the prediction.

16. One-way ANOVA because there are three independent groups, one variable, random sampling, normal distribution (all groups, N is greater than or equal to 30), and interval data homogeneous variance.

17. Single-sample t test because there is a single sample design in which the t assumptions of normality (16 subjects from a normally distributed population), random sampling, and interval or better data are met, but we do not know the population standard deviation.

18. Single-sample z test because we have both the population mean and standard deviation, a sample mean, and random sampling of a normally distributed population ($n = 30$).

19. Independent t test because there are two independent groups in which the independent t test assumptions of normality ($n = 35$ for both groups), interval or better data (number of problems correct), random sampling, and homogeneous variances are met.

20. Paired t test because we have before and after conditions with the same subjects participating in both conditions. We have a normal distribution because our n was greater that 30, and the data were measured on a ratio scale.

21. $Y' = b_y X + a_y$

22. This means that the variability around the regression line is uniform for all of the values of X.

23. This means that the variability around the regression line is NOT uniform for all of the values of X. For example, if the scores show much more variability at higher values of X, then these would be nonuniform data, and this breaks a fundamental assumption in interpreting the results of linear regression analyses, just as skewed data alter the interpretation of means and standard deviations.

CHAPTER 13 HOMEWORK ANSWERS

1. You can understand and interpret why there is an F-obtained value in your linear regression output. You can mix and match the measurement scales of your independent variables (this also allows you to analyze interactions among ordinal/nominal and interval/ratio variables).

2. Analysis of covariance

3. ANCOVA is a specialized form of ANOVA that is used when an investigator wishes to remove the effects of a variable that is known to influence the dependent variable but is not the subject of the current experiment and analysis.

4. They are all based on a general linear equation where there are multiple influences on any given score. The effects of the grand mean, the independent variable, and the unexplained variance (error) are all included in each conceptual formula.

5. ANOVA, regression, and t tests were developed independently of one another, and thus their terminology varies historically. The realization that all of these techniques are really the same and have the same underlying linear (additive) equations is more recent, so the traditional terminology associated with each test independent of the other tests is still widely prevalent in the literature.

6. All of these formulas share in common similar assumptions (i.e., normal distribution) and a linear form where the effects of each variable on the score are additive, or each term is added up.

7. They can all be solved using the same process (matrix algebra).

8. C. Categorical and continuous

CHAPTER 14 HOMEWORK ANSWERS

1. You are likely to reject the null hypothesis.

2. The scores in each cell are independent, and no f_e is less than 5 if the number of rows or columns is greater than 2.

3. The dependent variable or the difference scores must be ordinal.

4. The dependent variable must be ordinal or better.

5. The dependent variable must be ordinal or better, and if you are using the chi-square table, then there must be at least five scores in each sample.

6. Used when one or both of the variables are ordinal and the relationship between the variables is linear.

7. You do not use the chi-square test if f_e is below 5. If the experiment involves a 2×2 contingency table and the f_e is below 5, then Fisher's exact probability test can be used.

8. A contingency table is a two-way chi-square. In other words, there are two factors with multiple levels rather than one factor, and you are testing the likelihood that the frequencies are randomly distributed among the cells in the table.

9. Independent t test (Mann-Whitney U test)

   Paired t test (Wilcoxon)
   ANOVA (Kruskal-Wallis)

10. A. frequencies

11. $\chi^2_{obtained} = 2.0906$, $df = 1$, $\chi^2_{critical} = 3.84$; fail to reject null hypothesis. Age and soda preference appear to be unrelated.

12. $\chi^2_{obtained} = 38.0953$, $df = 2$, $\chi^2_{critical} = 5.99$; reject null hypothesis. Gender and watch preference appear to be related.

13. $\chi^2_{obtained} = 8.0000$, $df = 1$, $\chi^2_{critical} = 6.64$; reject null hypothesis. The distribution of opinions is not random.

14. $\chi^2_{obtained} = 53.662491$, $\chi^2_{critical} = 5.99$; reject null hypothesis. The distribution of opinions is not random. Conclude that reading performance has changed.

15. Kruskal-Wallis test because there are three independent groups, one variable, and random sampling, but the ANOVA assumptions are not met. The data are not normally distributed because n is not greater than or equal to 30 ($n = 3$) and we were not told to assume normality.

16. Wilcoxon-signed ranks test because we have a paired design in which the t test assumptions are not met. Although the data are interval or better (heartbeats per minute), we do not have a normal distribution (n is not greater than or equal to 30). The Wilcoxon signed ranks test was chosen over the sign test because it is more powerful.

17. Mann-Whitney U test because it is an independent groups design with two conditions and doesn't meet the assumptions for the independent t test. Although the ratio scaled data are randomly selected from a normally distributed population, the variances are not homogeneous.

18. Wilcoxon signed ranks test because there are two correlated groups in which the subject's anxiety after treatment is compared to the subject's baseline anxiety. The assumptions of normality and random sampling are met. The measurement, however, is ordinal.

19. Null: There is no relationship between pubertal timing and gender. Alternative: There is a relationship between pubertal timing and gender.

20. Chi-square critical = 5.99

21.

$$f_e = \frac{\text{row total} \times \text{column total}}{N} \qquad N = 368$$

Cell 1: 191 × 67/368 = 34.77

Cell 2: 191 × 60/368 = 31.14

Cell 3: 191 × 241/368 = 125.08

Cell 4: 177 × 67/368 = 32.23

Cell 5: 177 × 60/368 = 28.86

Cell 6: 177 × 241/368 = 115.92

22. Chi-square obtained = sum of $\dfrac{(f_o - f_e)^2}{f_e}$

    Cell 1: $(24 - 34.77)^2 / 34.77 = 3.34$

    Cell 2: $(39 - 31.14)^2 / 31.14 = 1.98$

    Cell 3: $(128 - 125.08)^2 / 125.08 = .07$

    Cell 4: $(43 - 32.23)^2 / 32.23 = 3.60$

    Cell 5: $(21 - 28.86)^2 / 28.86 = 2.14$

    Cell 6: $(113 - 115.92)^2 / 115.92 = .07$

    Sum of chi-square for each cell = chi-square obtained = 3.34 + 1.98 + .07 + 3.60 + 2.14 + .07 = 11.20

23. 11.20 is greater than or equal to 5.99, so reject null. There is a relationship between pubertal timing and gender.

CHAPTER 15 HOMEWORK ANSWERS

1. Mann-Whitney U test because the homogeneity of variance assumption is not met.

2. Wilcoxon signed ranks test because you cannot assume normality and the measurement scale is ordinal.

3. Independent t test because you have two groups and the assumptions are met.

4. Kruskal-Wallis test because there are four wines, but the ratings are ordinal.

5. Single-sample z test because you know the population mean and standard deviation.

6. Paired t test because the design is paired/correlated and you can assume normality based on the sample size.

7. Contingency table chi-square because frequencies are the dependent variable and there are two factors.

8. ANOVA because you have three groups and the assumptions are met.

9. Mann-Whitney U test because you cannot assume normality.

10. Paired t test because the subjects are matched individually with similar players and you can assume normality based on sample size.

11. Contingency table chi-square because you have two factors and the dependent variable is frequencies.

12. Chi-square because the data are frequencies (nominal) and you have one factor.

13. ANOVA because the assumptions are met and you have three groups.

14. Paired t test because you can compare lemon preference before and after conditioning on a large sample size, so you can assume normality.

15. Wilcoxon signed rank test because the sample size is unknown and you cannot assume normality.

Appendix

Statistical Tables

TABLE A. AREAS UNDER THE NORMAL DISTRIBUTION CURVE

Z	Area Between Mean and Z	Area Beyond Z	Z	Area Between Mean and Z	Area Beyond Z
A	B	C	A	B	C
0.00	0.0000	0.5000	0.19	0.0753	0.4247
0.01	0.0040	0.4960	0.20	0.0793	0.4207
0.02	0.0080	0.4920	0.21	0.0832	0.4168
0.03	0.0120	0.4880	0.22	0.0871	0.4129
0.04	0.0160	0.4840	0.23	0.0910	0.4090
0.05	0.0199	0.4801	0.24	0.0948	0.4052
0.06	0.0239	0.4761	0.25	0.0987	0.4013
0.07	0.0279	0.4721	0.26	0.1026	0.3974
0.08	0.0319	0.4681	0.27	0.1064	0.3936
0.09	0.0359	0.4641	0.28	0.1103	0.3897
0.10	0.0398	0.4602	0.29	0.1141	0.3859
0.11	0.0438	0.4562	0.30	0.1179	0.3821
0.12	0.0478	0.4522	0.31	0.1217	0.3783
0.13	0.0517	0.4483	0.32	0.1255	0.3745
0.14	0.0557	0.4443	0.33	0.1293	0.3707
0.15	0.0596	0.4404	0.34	0.1331	0.3669
0.16	0.0636	0.4364	0.35	0.1368	0.3632
0.17	0.0675	0.4325	0.36	0.1406	0.3594
0.18	0.0714	0.4286	0.37	0.1443	0.3557

(Continued)

(Continued)

Z	Area Between Mean and Z	Area Beyond Z	Z	Area Between Mean and Z	Area Beyond Z
A	B	C	A	B	C
0.38	0.1480	0.3520	0.68	0.2517	0.2483
0.39	0.1517	0.3483	0.69	0.2549	0.2451
0.40	0.1554	0.3446	0.70	0.2580	0.2420
0.41	0.1591	0.3409	0.71	0.2611	0.2389
0.42	0.1628	0.3372	0.72	0.2642	0.2358
0.43	0.1664	0.3336	0.73	0.2673	0.2327
0.44	0.1700	0.3300	0.74	0.2704	0.2296
0.45	0.1736	0.3264	0.75	0.2734	0.2266
0.46	0.1772	0.3228	0.76	0.2764	0.2236
0.47	0.1808	0.3192	0.77	0.2794	0.2206
0.48	0.1844	0.3156	0.78	0.2823	0.2177
0.49	0.1879	0.3121	0.79	0.2852	0.2148
0.50	0.1915	0.3085	0.80	0.2881	0.2119
0.51	0.1950	0.3050	0.81	0.2910	0.2090
0.52	0.1985	0.3015	0.82	0.2939	0.2061
0.53	0.2019	0.2981	0.83	0.2967	0.2033
0.54	0.2054	0.2946	0.84	0.2995	0.2005
0.55	0.2088	0.2912	0.85	0.3023	0.1977
0.56	0.2123	0.2877	0.86	0.3051	0.1949
0.57	0.2157	0.2843	0.87	0.3078	0.1922
0.58	0.2190	0.2810	0.88	0.3106	0.1894
0.59	0.2224	0.2776	0.89	0.3133	0.1867
0.60	0.2257	0.2743	0.90	0.3159	0.1841
0.61	0.2291	0.2709	0.91	0.3186	0.1814
0.62	0.2324	0.2676	0.92	0.3212	0.1788
0.63	0.2357	0.2643	0.93	0.3238	0.1762
0.64	0.2389	0.2611	0.94	0.3264	0.1736
0.65	0.2422	0.2578	0.95	0.3289	0.1711
0.66	0.2454	0.2546	0.96	0.3315	0.1685
0.67	0.2486	0.2514	0.97	0.3340	0.1660

Z	Area Between Mean and Z	Area Beyond Z	Z	Area Between Mean and Z	Area Beyond Z
A	B	C	A	B	C
0.98	0.3365	0.1635	1.29	0.4015	0.0985
0.99	0.3389	0.1611	1.30	0.4032	0.0968
1.00	0.3413	0.1587	1.31	0.4049	0.0951
1.01	0.3438	0.1562	1.32	0.4066	0.0934
1.02	0.3461	0.1539	1.33	0.4082	0.0918
1.03	0.3485	0.1515	1.34	0.4099	0.0901
1.04	0.3508	0.1492	1.35	0.4115	0.0885
1.05	0.3531	0.1469	1.36	0.4131	0.0869
1.06	0.3554	0.1446	1.37	0.4147	0.0853
1.07	0.3577	0.1423	1.38	0.4162	0.0838
1.08	0.3599	0.1401	1.39	0.4177	0.0823
1.09	0.3621	0.1379	1.40	0.4192	0.0808
1.10	0.3643	0.1357	1.41	0.4207	0.0793
1.11	0.3665	0.1335	1.42	0.4222	0.0778
1.12	0.3686	0.1314	1.43	0.4236	0.0764
1.13	0.3708	0.1292	1.44	0.4251	0.0749
1.14	0.3729	0.1271	1.45	0.4265	0.0735
1.15	0.3749	0.1251	1.46	0.4279	0.0721
1.16	0.3770	0.1230	1.47	0.4292	0.0708
1.17	0.3790	0.1210	1.48	0.4306	0.0694
1.18	0.3810	0.1190	1.49	0.4319	0.0681
1.19	0.3830	0.1170	1.50	0.4332	0.0668
1.20	0.3849	0.1151	1.51	0.4345	0.0655
1.21	0.3869	0.1131	1.52	0.4357	0.0643
1.22	0.3888	0.1112	1.53	0.4370	0.0630
1.23	0.3907	0.1093	1.54	0.4382	0.0618
1.24	0.3925	0.1075	1.55	0.4394	0.0606
1.25	0.3944	0.1056	1.56	0.4406	0.0594
1.26	0.3962	0.1038	1.57	0.4418	0.0582
1.27	0.3980	0.1020	1.58	0.4429	0.0571
1.28	0.3997	0.1003	1.59	0.4441	0.0559

(Continued)

(Continued)

Z	Area Between Mean and Z	Area Beyond Z	Z	Area Between Mean and Z	Area Beyond Z
A	B	C	A	B	C
1.60	0.4452	0.0548	1.89	0.4706	0.0294
1.61	0.4463	0.0537	1.90	0.4713	0.0287
1.62	0.4474	0.0526	1.91	0.4719	0.0281
1.63	0.4484	0.0516	1.92	0.4726	0.0274
1.64	0.4495	0.0505	1.93	0.4732	0.0268
1.65	0.4505	0.0495	1.94	0.4738	0.0262
1.66	0.4515	0.0485	1.95	0.4744	0.0256
1.67	0.4525	0.0475	1.96	0.4750	0.0250
1.68	0.4535	0.0465	1.97	0.4756	0.0244
1.69	0.4545	0.0455	1.98	0.4761	0.0239
1.70	0.4554	0.0446	1.99	0.4767	0.0233
1.71	0.4564	0.0436	2.00	0.4772	0.0228
1.72	0.4573	0.0427	2.01	0.4778	0.0222
1.73	0.4582	0.0418	2.02	0.4783	0.0217
1.74	0.4591	0.0409	2.03	0.4788	0.0212
1.75	0.4599	0.0401	2.04	0.4793	0.0207
1.76	0.4608	0.0392	2.05	0.4798	0.0202
1.77	0.4616	0.0384	2.06	0.4803	0.0197
1.78	0.4625	0.0375	2.07	0.4808	0.0192
1.79	0.4633	0.0367	2.08	0.4812	0.0188
1.80	0.4641	0.0359	2.09	0.4817	0.0183
1.81	0.4649	0.0351	2.10	0.4821	0.0179
1.82	0.4656	0.0344	2.11	0.4826	0.0174
1.83	0.4664	0.0336	2.12	0.4830	0.0170
1.84	0.4671	0.0329	2.13	0.4834	0.0166
1.85	0.4678	0.0322	2.14	0.4838	0.0162
1.86	0.4686	0.0314	2.15	0.4842	0.0158
1.87	0.4693	0.0307	2.16	0.4846	0.0154
1.88	0.4699	0.0301	2.17	0.4850	0.0150

Z	Area Between Mean and Z	Area Beyond Z	Z	Area Between Mean and Z	Area Beyond Z
A	B	C	A	B	C
2.18	0.4854	0.0146	2.47	0.4932	0.0068
2.19	0.4857	0.0143	2.48	0.4934	0.0066
2.20	0.4861	0.0139	2.49	0.4936	0.0064
2.21	0.4864	0.0136	2.50	0.4938	0.0062
2.22	0.4868	0.0132	2.51	0.4940	0.0060
2.23	0.4871	0.0129	2.52	0.4941	0.0059
2.24	0.4875	0.0125	2.53	0.4943	0.0057
2.25	0.4878	0.0122	2.54	0.4945	0.0055
2.26	0.4881	0.0119	2.55	0.4946	0.0054
2.27	0.4884	0.0116	2.56	0.4948	0.0052
2.28	0.4887	0.0113	2.57	0.4949	0.0051
2.29	0.4890	0.0110	2.58	0.4951	0.0049
2.30	0.4893	0.0107	2.59	0.4952	0.0048
2.31	0.4896	0.0104	2.60	0.4953	0.0047
2.32	0.4898	0.0102	2.61	0.4955	0.0045
2.33	0.4901	0.0099	2.62	0.4956	0.0044
2.34	0.4904	0.0096	2.63	0.4957	0.0043
2.35	0.4906	0.0094	2.64	0.4959	0.0041
2.36	0.4909	0.0091	2.65	0.4960	0.0040
2.37	0.4911	0.0089	2.66	0.4961	0.0039
2.38	0.4913	0.0087	2.67	0.4962	0.0038
2.39	0.4916	0.0084	2.68	0.4963	0.0037
2.40	0.4918	0.0082	2.69	0.4964	0.0036
2.41	0.4920	0.0080	2.70	0.4965	0.0035
2.42	0.4922	0.0078	2.71	0.4966	0.0034
2.43	0.4925	0.0075	2.72	0.4967	0.0033
2.44	0.4927	0.0073	2.73	0.4968	0.0032
2.45	0.4929	0.0071	2.74	0.4969	0.0031
2.46	0.4931	0.0069	2.75	0.4970	0.0030

(Continued)

(Continued)

Z	Area Between Mean and Z	Area Beyond Z	Z	Area Between Mean and Z	Area Beyond Z
A	B	C	A	B	C
2.76	0.4971	0.0029	3.05	0.4989	0.0011
2.77	0.4972	0.0028	3.06	0.4989	0.0011
2.78	0.4973	0.0027	3.07	0.4989	0.0011
2.79	0.4974	0.0026	3.08	0.4990	0.0010
2.80	0.4974	0.0026	3.09	0.4990	0.0010
2.81	0.4975	0.0025	3.10	0.4990	0.0010
2.82	0.4976	0.0024	3.11	0.4991	0.0009
2.83	0.4977	0.0023	3.12	0.4991	0.0009
2.84	0.4977	0.0023	3.13	0.4991	0.0009
2.85	0.4978	0.0022	3.14	0.4992	0.0008
2.86	0.4979	0.0021	3.15	0.4992	0.0008
2.87	0.4979	0.0021	3.16	0.4992	0.0008
2.88	0.4980	0.0020	3.17	0.4992	0.0008
2.89	0.4981	0.0019	3.18	0.4993	0.0007
2.90	0.4981	0.0019	3.19	0.4993	0.0007
2.91	0.4982	0.0018	3.20	0.4993	0.0007
2.92	0.4982	0.0018	3.21	0.4993	0.0007
2.93	0.4983	0.0017	3.22	0.4994	0.0006
2.94	0.4984	0.0016	3.23	0.4994	0.0006
2.95	0.4984	0.0016	3.24	0.4994	0.0006
2.96	0.4985	0.0015	3.25	0.4994	0.0006
2.97	0.4985	0.0015	3.26	0.4994	0.0006
2.98	0.4986	0.0014	3.27	0.4995	0.0005
2.99	0.4986	0.0014	3.28	0.4995	0.0005
3.00	0.4987	0.0013	3.29	0.4995	0.0005
3.01	0.4987	0.0013	3.30	0.4995	0.0005
3.02	0.4987	0.0013	3.31	0.4995	0.0005
3.03	0.4988	0.0012	3.32	0.4995	0.0005
3.04	0.4988	0.0012	3.33	0.4996	0.0004

Z	Area Between Mean and Z	Area Beyond Z	Z	Area Between Mean and Z	Area Beyond Z
A	B	C	A	B	C
3.34	0.4996	0.0004	3.53	0.4998	0.0002
3.35	0.4996	0.0004	3.54	0.4998	0.0002
3.36	0.4996	0.0004	3.55	0.4998	0.0002
3.37	0.4996	0.0004	3.56	0.4998	0.0002
3.38	0.4996	0.0004	3.57	0.4998	0.0002
3.39	0.4997	0.0003	3.58	0.4998	0.0002
3.40	0.4997	0.0003	3.59	0.4998	0.0002
3.41	0.4997	0.0003	3.60	0.4998	0.0002
3.42	0.4997	0.0003	3.61	0.4998	0.0002
3.43	0.4997	0.0003	3.62	0.4999	0.0001
3.44	0.4997	0.0003	3.63	0.4999	0.0001
3.45	0.4997	0.0003	3.64	0.4999	0.0001
3.46	0.4997	0.0003	3.65	0.4999	0.0001
3.47	0.4997	0.0003	3.66	0.4999	0.0001
3.48	0.4997	0.0003	3.67	0.4999	0.0001
3.49	0.4998	0.0002	3.68	0.4999	0.0001
3.50	0.4998	0.0002	3.69	0.4999	0.0001
3.51	0.4998	0.0002	3.70	0.4999	0.0001
3.52	0.4998	0.0002			

TABLE B. CRITICAL VALUES OF THE STUDENT T DISTRIBUTION

	Level of Significance for One-Tailed Test					
	0.10	0.05	0.025	0.010	0.005	0.0005
	Level of Significance for Two-Tailed Test					
Df	0.20	0.10	0.05	0.02	0.01	0.001
1	3.078	6.314	12.706	31.821	63.657	636.619
2	1.886	2.920	4.303	6.965	9.925	31.599
3	1.638	2.353	3.182	4.541	5.841	12.924
4	1.533	2.132	2.776	3.747	4.604	8.610
5	1.476	2.015	2.571	3.365	4.032	6.869
6	1.440	1.943	2.447	3.143	3.707	5.959
7	1.415	1.895	2.365	2.998	3.499	5.408
8	1.397	1.860	2.306	2.896	3.355	5.041
9	1.383	1.833	2.262	2.821	3.250	4.781
10	1.372	1.812	2.228	2.764	3.169	4.587
11	1.363	1.796	2.201	2.718	3.106	4.437
12	1.356	1.782	2.179	2.681	3.055	4.318
13	1.350	1.771	2.160	2.650	3.012	4.221
14	1.345	1.761	2.145	2.624	2.977	4.140
15	1.341	1.753	2.131	2.602	2.947	4.073
16	1.337	1.746	2.120	2.583	2.921	4.015
17	1.333	1.740	2.110	2.567	2.898	3.965
18	1.330	1.734	2.101	2.552	2.878	3.922
19	1.328	1.729	2.093	2.539	2.861	3.883
20	1.325	1.725	2.086	2.528	2.845	3.850
21	1.323	1.721	2.080	2.518	2.831	3.819
22	1.321	1.717	2.074	2.508	2.819	3.792
23	1.319	1.714	2.069	2.500	2.807	3.768
24	1.318	1.711	2.064	2.492	2.797	3.745
25	1.316	1.708	2.060	2.485	2.787	3.725

Df	Level of Significance for One-Tailed Test					
	0.10	0.05	0.025	0.010	0.005	0.0005
	Level of Significance for Two-Tailed Test					
	0.20	0.10	0.05	0.02	0.01	0.001
28	1.313	1.701	2.048	2.467	2.763	3.674
29	1.311	1.699	2.045	2.462	2.756	3.659
30	1.310	1.697	2.042	2.457	2.750	3.646
40	1.303	1.684	2.021	2.423	2.704	3.551
60	1.296	1.671	2.000	2.390	2.660	3.460
120	1.289	1.658	1.980	2.358	2.617	3.373
∞	1.282	1.645	1.960	2.327	2.576	3.291

TABLE C. *F* DISTRIBUTION

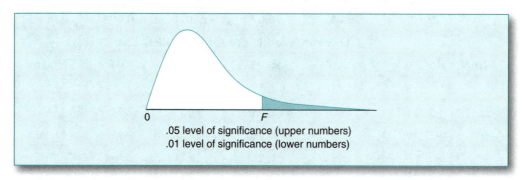

0   F
.05 level of significance (upper numbers)
.01 level of significance (lower numbers)

df$_2$	df$_1$									
	1	2	3	4	5	6	7	8	9	10
11	4.84	3.98	3.59	3.36	3.20	3.09	3.01	2.95	2.90	2.86
	9.65	7.20	6.22	5.67	5.32	5.07	4.88	4.74	4.63	4.54
12	4.75	3.88	3.49	3.26	3.11	3.00	2.92	2.85	2.80	2.76
	9.33	6.93	5.95	5.41	5.06	4.82	4.65	4.50	4.39	4.30
13	4.67	3.80	3.41	3.18	3.02	2.92	2.84	2.77	2.72	2.67
	9.07	6.70	5.74	5.20	4.86	4.62	4.44	4.30	4.19	4.10
14	4.60	3.74	3.34	3.11	2.96	2.85	2.77	2.70	2.65	2.60
	8.86	6.51	5.56	5.83	4.69	4.46	4.28	4.14	4.03	3.94
15	4.54	3.68	3.29	3.06	2.90	2.79	2.70	2.64	2.59	2.55
	8.68	6.36	5.42	4.89	4.56	4.32	4.14	4.00	3.89	3.80
16	4.49	3.63	3.24	3.01	2.85	2.74	2.66	2.59	2.54	2.49
	8.53	6.23	5.29	4.77	4.44	4.28	4.03	3.89	3.78	3.69
17	4.45	3.59	3.20	2.96	2.81	2.70	2.62	2.55	2.50	2.45
	8.40	6.11	5.18	4.67	4.34	4.10	3.93	3.79	3.68	3.59
18	4.41	3.55	3.16	2.93	2.77	2.66	2.58	2.51	2.46	2.41
	8.28	6.01	5.08	4.58	4.25	4.01	3.85	3.71	3.60	3.51
19	4.38	3.52	3.13	2.90	2.74	2.63	2.55	2.48	2.43	2.38
	8.18	5.83	5.01	4.50	4.17	3.94	3.77	3.63	3.52	3.43
20	4.35	3.49	3.10	2.87	2.71	2.60	2.52	2.45	2.40	2.35
	8.10	5.85	4.94	4.43	4.10	3.87	3.71	3.56	3.45	3.37
21	4.32	3.47	3.07	2.84	2.68	2.57	2.49	2.42	2.37	2.32
	8.02	5.78	4.87	4.37	4.04	3.81	3.65	3.51	3.40	3.31

df$_2$	\multicolumn{10}{c}{df$_1$}
	1	**2**	**3**	**4**	**5**	**6**	**7**	**8**	**9**	**10**
22	4.30	3.44	3.05	2.82	2.66	2.55	2.47	2.40	2.35	2.30
	7.94	5.72	4.82	4.31	3.89	3.76	3.59	3.45	3.35	3.26
23	4.28	3.42	3.03	2.80	2.64	2.53	2.45	2.38	2.32	2.28
	7.88	5.66	4.76	4.26	3.94	3.71	3.54	3.41	3.30	3.21
27	4.21	3.35	2.96	2.73	2.57	2.46	2.37	2.30	2.25	2.20
	7.68	5.49	4.60	4.11	3.79	3.56	3.39	3.26	3.14	3.06
28	4.20	3.34	2.95	2.71	2.56	2.44	2.36	2.29	2.24	2.19
	7.64	5.45	4.57	4.07	3.76	3.53	3.36	3.23	3.11	3.03
29	4.18	3.33	2.93	2.70	2.54	2.43	2.35	2.28	2.22	2.18
	7.60	5.42	4.54	4.04	3.73	3.50	3.33	3.20	3.08	3.00
30	4.17	3.32	2.92	2.69	2.53	2.42	2.34	2.27	2.21	2.16
	7.56	5.39	4.51	4.02	3.70	3.47	3.30	3.17	3.06	2.98
32	4.15	3.30	2.90	2.67	2.51	2.40	2.32	2.25	2.19	2.14
	7.50	5.34	4.46	3.97	3.66	3.42	3.25	3.12	3.01	2.94
34	4.13	3.28	2.88	2.65	2.49	2.38	2.30	2.23	2.17	2.12
	7.44	5.29	4.42	3.93	3.61	3.38	3.21	3.08	2.97	2.89
36	4.11	3.26	2.86	2.63	2.48	2.36	2.28	2.21	2.15	2.10
	7.39	5.25	4.38	3.89	3.58	3.35	3.18	3.04	2.94	2.86
38	4.10	3.25	2.85	2.62	2.46	2.35	2.26	2.19	2.14	2.09
	7.35	5.21	4.34	3.86	3.54	3.32	3.15	3.02	2.91	2.82
40	4.08	3.23	2.84	2.61	2.45	2.34	2.25	2.18	2.12	2.07
	7.31	5.18	4.31	3.83	3.51	3.29	3.12	2.99	2.88	2.80
42	4.07	3.22	2.83	2.59	2.44	2.32	2.24	2.17	2.11	2.06
	7.27	5.15	4.29	3.80	3.49	3.26	3.10	2.96	2.86	2.77
44	4.06	3.21	2.82	2.58	2.43	2.31	2.23	2.16	2.10	2.05
	7.24	5.12	4.26	3.78	3.46	3.24	3.07	2.94	2.84	2.75
46	4.05	3.20	2.81	2.57	2.42	2.30	2.22	2.14	2.09	2.04
	7.21	5.10	4.24	3.76	3.44	3.22	3.05	2.92	2.82	2.73
48	4.04	3.19	2.80	2.56	2.41	2.30	2.21	2.14	2.08	2.03
	7.19	5.08	4.22	3.74	3.42	3.20	3.04	2.90	2.80	2.71

(Continued)

(Continued)

df_2	df_1									
	1	2	3	4	5	6	7	8	9	10
50	4.03	3.18	2.79	2.56	2.40	2.29	2.20	2.13	2.07	2.02
	7.17	5.06	4.20	3.72	3.41	3.18	3.02	2.88	2.78	2.70
55	4.02	3.17	2.78	2.54	2.38	2.27	2.18	2.11	2.05	2.00
	7.12	5.01	4.16	3.68	3.37	3.15	2.98	2.85	2.75	2.66
60	4.00	3.15	2.76	2.52	2.37	2.25	2.17	2.10	2.04	1.99
	7.08	4.98	4.13	3.65	3.34	3.12	2.95	2.82	2.72	2.63
65	3.99	3.14	2.75	2.51	2.36	2.24	2.15	2.08	2.02	1.98
	7.04	4.95	4.10	3.62	3.31	3.09	2.93	2.79	2.70	2.61
70	3.98	3.13	2.74	2.50	2.35	2.23	2.14	2.07	2.01	1.97
	7.01	4.92	4.08	3.60	3.29	3.07	2.91	2.77	2.67	2.59
80	3.96	3.11	2.72	2.48	2.33	2.21	2.12	2.05	1.99	1.95
	6.96	4.88	4.04	3.56	3.25	3.04	2.87	2.74	2.64	2.55
100	3.94	3.09	2.70	2.46	2.30	2.19	2.10	2.03	1.97	1.92
	6.90	4.82	3.98	3.51	3.20	2.99	2.82	2.69	2.59	2.51
125	3.92	3.07	2.68	2.44	2.29	2.17	2.08	2.01	1.95	1.90
	6.84	4.78	3.94	3.47	3.17	2.95	2.79	2.65	2.56	2.47
150	3.91	3.06	2.67	2.43	2.27	2.16	2.07	2.00	1.94	1.89
	6.81	4.75	3.91	3.44	3.14	2.92	2.76	2.62	2.53	2.44
200	3.89	3.04	2.65	2.41	2.26	2.14	2.05	1.98	1.92	1.87
	6.76	4.71	3.88	3.41	3.11	2.90	2.73	2.60	2.50	2.41

Note: Reject the null hypothesis if the derived F value is equal to or greater than the tabled F value. When $df_1 = \infty$ and $df_2 = \infty$, the tabled critical value of $F = 1.00$ at $p = .05$ and $.01$.

TABLE D. TABLE OF CRITICAL Q-VALUES (FOR TUKEY HSD)

df_{with}	k = Number of Groups									
	α	2	3	4	5	6	7	8	9	10
5	.05	3.64	4.60	5.22	5.67	6.03	6.33	6.58	6.80	6.99
	.01	5.70	6.98	7.80	8.42	8.91	9.32	9.67	9.97	10.24
6	.05	3.46	4.34	4.90	5.30	5.63	5.90	6.12	6.32	6.49
	.01	5.24	6.33	7.03	7.56	7.97	8.32	8.61	8.87	9.10
7	.05	3.34	4.16	4.68	5.06	5.36	5.61	5.82	6.00	6.16
	.01	4.95	5.92	6.54	7.01	7.37	7.68	7.94	8.17	8.37
8	.05	3.26	4.04	4.53	4.89	5.17	5.40	5.60	5.77	5.92
	.01	4.75	5.64	6.20	6.62	6.96	7.24	7.47	7.68	7.86
9	.05	3.20	3.95	4.41	4.76	5.02	5.24	5.43	5.59	5.74
	.01	4.60	5.43	5.96	6.35	6.66	6.91	7.13	7.33	7.49
10	.05	3.15	3.88	4.33	4.65	4.91	5.12	5.30	5.46	5.60
	.01	4.48	5.27	5.77	6.14	6.43	6.67	6.87	7.05	7.21
11	.05	3.11	3.82	4.26	4.57	4.82	5.03	5.20	5.35	5.49
	.01	4.39	5.15	5.62	5.97	6.25	6.48	6.67	6.84	6.99
12	.05	3.08	3.77	4.20	4.51	4.75	4.95	5.12	5.27	5.39
	.01	4.32	5.05	5.50	5.84	6.10	6.32	6.51	6.67	6.81
13	.05	3.06	3.73	4.15	4.45	4.69	4.88	5.05	5.19	5.32
	.01	4.26	4.96	5.40	5.73	5.98	6.19	6.37	6.53	6.67
14	.05	3.03	3.70	4.11	4.41	4.64	4.83	4.99	5.13	5.25
	.01	4.21	4.89	5.32	5.63	5.88	6.08	6.26	6.41	6.54
15	.05	3.01	3.67	4.08	4.37	4.59	4.78	4.94	5.08	5.20
	.01	4.17	4.84	5.25	5.56	5.80	5.99	6.16	6.31	6.44
16	.05	3.00	3.65	4.05	4.33	4.56	4.74	4.90	5.03	5.15
	.01	4.13	4.79	5.19	5.49	5.72	5.92	6.08	6.22	6.35
17	.05	2.98	3.63	4.02	4.30	4.52	4.70	4.86	4.99	5.11
	.01	4.10	4.74	5.14	5.43	5.66	5.85	6.01	6.15	6.27

(Continued)

(Continued)

df_{with}	α	k = Number of Groups								
		2	3	4	5	6	7	8	9	10
18	.05	2.97	3.61	4.00	4.28	4.49	4.67	4.82	4.96	5.07
	.01	4.07	4.70	5.09	5.38	5.60	5.79	5.94	6.08	6.20
19	.05	2.96	3.59	3.98	4.25	4.47	4.65	4.79	4.92	5.04
	.01	4.05	4.67	5.05	5.33	5.55	5.73	5.89	6.02	6.14
20	.05	2.95	3.58	3.96	4.23	4.45	4.62	4.77	4.90	5.01
	.01	4.02	4.64	5.02	5.29	5.51	5.69	5.84	5.97	6.09
24	.05	2.92	3.53	3.90	4.17	4.37	4.54	4.68	4.81	4.92
	.01	3.96	4.55	4.91	5.17	5.37	5.54	5.69	5.81	5.92
30	.05	2.89	3.49	3.85	4.10	4.30	4.46	4.60	4.72	4.82
	.01	3.89	4.45	4.80	5.05	5.24	5.40	5.54	5.65	5.76
40	.05	2.86	3.44	3.79	4.04	4.23	4.39	4.52	4.63	4.73
	.01	3.82	4.37	4.70	4.93	5.11	5.26	5.39	5.50	5.60
60	.05	2.83	3.40	3.74	3.98	4.16	4.31	4.44	4.55	4.65
	.01	3.76	4.28	4.59	4.82	4.99	5.13	5.25	5.36	5.45
120	.05	2.80	3.36	3.68	3.92	4.10	4.24	4.36	4.47	4.56
	.01	3.70	4.20	4.50	4.71	4.87	5.01	5.12	5.21	5.30
∞	.05	2.77	3.31	3.63	3.86	4.03	4.17	4.29	4.39	4.47
	.01	3.64	4.12	4.40	4.60	4.76	4.88	4.99	5.08	5.16

TABLE E. PEARSON'S *r*

Critical Values of Pearson's *r*

	α Level of Significance			
One-Tailed:	**0.05**	**0.025**	**0.005**	**0.0005**
Two-Tailed:	**0.1**	**0.05**	**0.01**	**0.001**
df*				
1	0.9877	0.9969	0.9999	1.0000
2	0.9000	0.9500	0.9900	0.9990
3	0.8054	0.8783	0.9587	0.9911
4	0.7293	0.8114	0.9172	0.9741
5	0.6694	0.7545	0.8745	0.9509
6	0.6215	0.7067	0.8343	0.9249
7	0.5822	0.6664	0.7977	0.8983
8	0.5494	0.6319	0.7646	0.8721
9	0.5214	0.6021	0.7348	0.8470
10	0.4973	0.5760	0.7079	0.8233
11	0.4762	0.5529	0.6835	0.8010
12	0.4575	0.5324	0.6614	0.7800
13	0.4409	0.5140	0.6411	0.7604
14	0.4259	0.4973	0.6226	0.7419
15	0.4124	0.4821	0.6055	0.7247
16	0.4000	0.4683	0.5897	0.7084
17	0.3887	0.4555	0.5751	0.6932
18	0.3783	0.4438	0.5614	0.6788
19	0.3687	0.4329	0.5487	0.6652
20	0.3598	0.4227	0.5368	0.6524
21	0.3515	0.4132	0.5256	0.6402
22	0.3438	0.4044	0.5151	0.6287
23	0.3365	0.3961	0.5052	0.6178
24	0.3297	0.3882	0.4958	0.6074
25	0.3233	0.3809	0.4869	0.5974

(Continued)

(Continued)

α Level of Significance			
One-Tailed: 0.05	0.025	0.005	0.0005
Two-Tailed: 0.1	0.05	0.01	0.001
df*			
26	0.3172	0.3739	0.4785	0.5880
27	0.3115	0.3673	0.4705	0.5790
28	0.3061	0.3610	0.4629	0.5703
29	0.3009	0.3550	0.4556	0.5620
30	0.2960	0.3494	0.4487	0.5541
40	0.2573	0.3044	0.3932	0.4896
50	0.2306	0.2732	0.3542	0.4432
100	0.1638	0.1946	0.2540	0.3211
150	0.1339	0.1593	0.2083	0.2643
200	0.1161	0.1381	0.1809	0.2298
500	0.0735	0.0875	0.1149	0.1464
1000	0.0520	0.0619	0.0813	0.1038
10000	0.0164	0.0196	0.0258	0.0329
100000	0.0052	0.0062	0.0081	0.0104

*Note: $df = N - 2$

TABLE F. THE CHI-SQUARE DISTRIBUTION

df	α Level of Significance			
	.05	.02	.01	.001
1	3.84	5.41	6.64	10.38
2	5.99	7.82	9.21	13.82
3	7.82	9.84	11.34	16.27
4	9.49	11.67	13.28	18.46
5	11.07	13.39	15.09	20.52
6	12.59	15.03	16.81	22.46
7	14.07	16.62	18.48	24.32
8	15.51	18.17	20.09	26.12
9	16.92	19.68	21.67	27.88
10	18.31	21.16	23.21	29.59
11	19.68	22.62	24.72	31.26
12	21.03	24.05	26.22	32.91
13	22.36	25.47	27.69	34.53
14	23.68	26.87	29.14	36.12
15	25.00	28.26	30.58	37.70
16	26.30	29.63	32.00	39.25
17	27.59	31.00	33.41	40.79
18	28.87	32.35	34.80	42.31
19	30.14	33.69	36.19	43.82
20	31.41	35.02	37.57	45.32
21	32.67	36.34	38.93	46.80
22	33.92	37.66	40.29	48.27
23	35.17	38.97	41.64	49.73
24	36.42	40.27	42.98	51.18
25	37.65	41.57	44.31	52.62
26	38.88	42.86	45.64	54.05
27	40.11	44.14	46.96	55.48
28	41.34	45.42	48.28	56.89

(Continued)

(Continued)

df	α Level of Significance			
	.05	.02	.01	.001
29	42.56	46.69	49.59	58.30
30	43.77	47.96	50.89	59.70

Source: This table is taken from Table IV of Fisher and Yates (1995), *Statistical Tables for Biological, Agricultural, and Medical Research,* published by Longman Group Ltd., London (previously published by Oliver and Boyd, Ltd., Edinburgh).

Note: Reject the null hypothesis if the derived chi-square value is equal to or greater than the tabled chi-square value.

TABLE G. CRITICAL VALUES (*T*) FOR WILCOXON'S SIGNED-RANKS TEST

To use this table: Compare your obtained value of Wilcoxon's statistic to the critical value in the table, for the given value of *N*. Your obtained value is statistically significant if it is equal to or *smaller* than the value in the table.

	One-Tailed Significance Levels		
	0.025	0.01	0.005
	Two-Tailed Significance Levels		
N	0.05	0.02	0.01
6	0	—	—
7	2	0	—
8	4	2	0
9	6	3	2
10	8	5	3
11	11	7	5
12	14	10	7
13	17	13	10
14	21	16	13
15	25	20	16
16	30	24	20
17	35	28	23
18	40	33	28
19	46	38	32
20	52	43	38
21	59	49	43
22	66	56	49
23	73	62	55
24	81	69	61
25	89	77	68

TABLE H. CRITICAL VALUES OF THE MANN-WHITNEY U

Two-Tailed Testing

n_2	α	3	4	5	6	7	8	9	10	11	12	13	14	15	16	17	18	19	20
3	.05	–	–	0	1	1	2	2	3	3	4	4	5	5	6	6	7	7	8
	.01	–	–	–	–	–	–	0	0	0	1	1	1	2	2	2	2	3	3
4	.05		0	1	2	3	4	4	5	6	7	8	9	10	11	11	12	13	14
	.01		–	–	0	0	1	1	2	2	3	3	4	5	5	6	6	7	8
5	.05			2	3	5	6	7	8	9	11	12	13	14	15	17	18	19	20
	.01			0	1	1	2	3	4	5	6	7	7	8	9	10	11	12	13
6	.05				5	6	8	10	11	13	14	16	17	19	21	22	24	25	27
	.01				2	3	4	5	6	7	9	10	11	12	13	15	16	17	18
7	.05					8	10	12	14	16	18	20	22	24	26	28	30	32	34
	.01					4	6	7	9	10	12	13	15	16	18	19	21	22	24
8	.05						13	15	17	19	22	24	26	29	31	34	36	38	41
	.01						7	9	11	13	15	17	18	20	22	24	26	28	30
9	.05							17	20	23	26	28	31	34	37	39	42	45	48
	.01							11	13	16	18	20	22	24	27	29	31	33	36
10	.05								23	26	29	33	36	39	42	45	48	52	55
	.01								16	18	21	24	26	29	31	34	37	39	42
11	.05									30	33	37	40	44	47	51	55	58	62
	.01									21	24	27	30	33	36	39	42	45	48
12	.05										37	41	45	49	53	57	61	65	69
	.01										27	31	34	37	41	44	47	51	54
13	.05											45	50	54	59	63	67	72	76
	.01											34	38	42	45	49	53	57	60
14	.05												55	59	64	69	74	78	83
	.01												42	46	50	54	58	63	67
15	.05													64	70	75	80	85	90
	.01													51	55	60	64	69	73
16	.05														75	81	86	92	98
	.01														60	65	70	74	79
17	.05															87	93	99	105
	.01															70	75	81	86
18	.05																99	106	112
	.01																81	87	92
19	.05																	113	119
	.01																	93	99
20	.05																		127
	.01																		105

CRITICAL VALUES OF THE MANN-WHITNEY U

One-Tailed Testing

n_2	α	3	4	5	6	7	8	9	10	11	12	13	14	15	16	17	18	19	20
3	.05	–	0	1	2	2	3	3	4	5	5	6	7	7	8	9	9	10	11
	.01	–	–	–	–	0	0	1	1	1	2	2	2	3	3	4	4	4	5
4	.05		1	2	3	4	5	6	7	8	9	10	11	12	14	15	16	17	18
	.01		–	0	1	1	2	3	3	4	5	5	6	7	7	8	9	9	10
5	.05			4	5	6	8	9	11	12	13	15	16	18	19	20	22	23	25
	.01			1	2	3	4	5	6	7	8	9	10	11	12	13	14	15	16
6	.05				7	8	10	12	14	16	17	19	21	23	25	26	28	30	32
	.01				3	4	6	7	8	9	11	12	13	15	16	18	19	20	22
7	.05					11	13	15	17	19	21	24	26	28	30	33	35	37	39
	.01					6	7	9	11	12	14	16	17	19	21	23	24	26	28
8	.05						15	18	20	23	26	28	31	33	36	39	41	44	47
	.01						9	11	13	15	17	20	22	24	26	28	30	32	34
9	.05							21	24	27	30	33	36	39	42	45	48	51	54
	.01							14	16	18	21	23	26	28	31	33	36	38	40
10	.05								27	31	34	37	41	44	48	51	55	58	62
	.01								19	22	24	27	30	33	36	38	41	44	47
11	.05									34	38	42	46	50	54	57	61	65	69
	.01									25	28	31	34	37	41	44	47	50	53
12	.05										42	47	51	55	60	64	68	72	77
	.01										31	35	38	42	46	49	53	56	60
13	.05											51	56	61	65	70	75	80	84
	.01											39	43	47	51	55	59	63	67
14	.05												61	66	71	77	82	87	92
	.01												47	51	56	60	65	69	73
15	.05													72	77	83	88	94	100
	.01													56	61	66	70	75	80
16	.05														83	89	95	101	107
	.01														66	71	76	82	87
17	.05															96	102	109	115
	.01															77	82	88	93
18	.05																109	116	123
	.01																88	94	100
19	.05																	123	130
	.01																	101	107
20	.05																		138
	.01																		114

TABLE I. SPEARMAN'S CORRELATION

Number of Pairs (n)	Level of Significance for a One-Tailed Test			
	.05	.025	.01	.005
	Level of Significance for a Two-Tailed Test			
	.10	.05	.02	.01
5	0.900	1.000	1.000	—
6	0.829	0.886	0.943	1.000
7	0.714	0.786	0.893	0.929
8	0.643	0.738	0.833	0.881
9	0.600	0.683	0.783	0.833
10	0.564	0.648	0.746	0.794
12	0.506	0.591	0.712	0.777
14	0.456	0.544	0.645	0.715
16	0.425	0.506	0.601	0.665
18	0.399	0.475	0.564	0.625
20	0.377	0.450	0.534	0.591
22	0.359	0.428	0.508	0.562
24	0.343	0.409	0.485	0.537
26	0.329	0.392	0.465	0.515
28	0.317	0.377	0.448	0.496
30	0.306	0.364	0.432	0.478

Note: Reject the null hypothesis if the derived Spearman coefficient is equal to or greater than the tabled Spearman coefficient value.

Glossary

Alternative hypothesis (H_A)   There is an effect or relationship between two variables.

Bar graph   Typically a two-dimensional graph wherein one dimension represents the independent variable and the other represents the dependent variable (frequency). The magnitude of the dependent variable is represented by the height of a rectangle with a uniform width (a bar).

Confidence interval for the population mean   Range of values with a calculated probability of containing the mean.

Confidence limits for the population mean   The upper and lower values (or boundaries) surrounding the confidence interval.

Constant   A value that is fixed.

Continuous variable   A variable that can have an infinite number of values between each unit.

Cumulative frequency   The sum of the frequency of scores starting from the lowest class interval and working up to the highest class interval. The cumulative frequency (or sum of the frequencies for each class interval) for the highest class interval should be equal to the sample size (N).

Cumulative percentage   The cumulative frequency divided by N and then the product multiplied by 100. The cumulative percentage for the highest class interval should equal 100.

Data   A measurement of a variable.

Degrees of freedom   The number of scores that are "free to vary" when calculating a statistic. The remaining value or values are then fixed.

Dependent variable   An outcome property that the researcher measures.

Descriptive statistics   Numbers that summarize a set of data in one of four ways, including a measure of central tendency, a measure of the variability of the scores, a measure of the shape of the distribution of scores, and a measure of the size of the sample.

Directional hypothesis (one-tailed)   The direction of the effect is indicated in the alternative hypothesis, and the null hypothesis suggests the opposite or no effect.

Discrete variable   A variable that has no values in between each unit.

Effect size   A measure or estimate of the strength of the relationship between two variables. It can be calculated for a population or estimated for a sample.

Empirical probability   The number of times an event classifiable as A occurred divided by the total number of occurrences.

Frequency (f)   The number of scores in that class interval.

Frequency curve   A two-dimensional graph plotting a continuous variable on the X-axis and frequency on the Y-axis with a smoothed line. Technically, it is a smoothed version of a histogram or graphical representation of a frequency distribution.

Frequency distribution   Data are collapsed into classes or ranges of values, and then the number of observations (or frequency) of each class, or range of values, is recorded in a table or graph.

Histogram   The most common form is plotted by splitting the data into equal-sized intervals (X-axis) and then indicating the frequency of scores in each interval (Y-axis) with a vertical bar.

Homoscedastic   Variance around line is uniform across all Xs.

Imperfect relationships   All data points do not fall on the best-fit line.

Independent variable   A predictor property that the researcher either manipulates experimentally or records as a naturally changing condition (thus, a quasi-independent variable).

Inferential statistics   Use of sample statistics in an attempt to infer the characteristics of a population of scores and ultimately to make decisions about experimental hypotheses.

Interaction   The combined effect of both (or all) independent variables over and above the separate effects of each variable alone. This is often reflected in the fact that one of your independent variables affects the dependent variable differently at different levels of your second independent variable.

Interval Data are collected in a manner that measures actual magnitude and has equal intervals between possible scores but does not have a meaningful absolute zero point.

Kurtosis A descriptive statistical measure of whether the frequency of scores in a distribution is more or less heavily clustered around the mean compared to a normal distribution. A distribution with a kurtosis = 0 is normally distributed and not kurtotic.

Main effect Influence of your independent variable on your dependent variable.

Mean A statistic that measures the middle, or central tendency, of a set of scores and is calculated by taking the sum of the scores and dividing by the number of scores.

Median The value at which 50% of the scores fall below it.

Mode The most frequent score in a set of scores.

Negative (inverse) relationships As X increases, Y decreases.

Nominal Data are on a categorical, and often qualitative, scale rather than one that is quantitative.

Nondirectional hypothesis (two-tailed) The direction of the effect is not indicated in the alternative hypothesis, and the null hypothesis suggests only that there is no effect or relationship between the variables.

Nonparametric tests Statistical tests that do not require assumptions about the underlying parameters of the population.

Null hypothesis (H$_0$) Any effect or relationship between two variables is due to chance factors.

Ordinal Data are on a categorical scale, in which categories can be ranked in relative order.

Parameter A summary calculation on data collected from a population.

Parametric tests Statistical tests require assumptions about the parameters of the population.

Perfect relationships All data points fall on the best-fit line.

Population The complete set of individuals, objects, or scores that is under study. The entire set of scores.

Positive (direct) relationships   As X increases, Y increases.

Power   A measure (between 0 and 1.00) of the ability of a statistical test to detect an effect of the independent variable if the effect exists in reality.

Probability   A number between 0 and 1 that indicates the likelihood of an event occurring due to chance factors alone.

Random sampling   Subjects are selected in a manner that ensures that each possible sample of a given size is equally likely to be selected and each subject in the population is equally likely to be selected for the sample.

Range   A statistic that measures the variability of a set of scores and is calculated by taking the largest score and subtracting the smallest score.

Ratio   Data are collected in a manner that measures magnitude, has equal intervals between possible scores, and contains an absolute zero point.

Real effect   When your independent variable affects your dependent variable.

Relative frequency   The frequency of scores in that class interval divided by the total number of scores in the distribution.

Sample   A subset of the population of interest.

Sampling distribution   A hypothetical representation of all possible values of a statistic with the probability of each possible value if no independent variable is acting.

Sampling distribution of the mean   A theoretical distribution of means obtained by drawing all possible samples of a given size from a population and calculating a mean for each sample. From this distribution, the probability of obtaining any given mean from a population with the same parameters by chance can be calculated.

Sampling with replacement   Each subject that is selected for a sample is returned to the pool prior to the next selection.

Sampling without replacement   Each selected subject does not return to the pool prior to the next selection.

s_B^2 Between-group variance (reflects natural variation in the population [E] and the effect of the independent variable, if any).

s_C^2 Between-column variance (reflects natural variation in the population and any variability due to your column variable).

Scatter plot   A graphical display of pairs of continuous data points (X and Y pairs) that is intended to reveal any potential relationship between the two variables.

Skewness   A descriptive statistical measure of whether a distribution has a higher frequency of scores on one side of the mean compared to the other side of the mean. A distribution with a skewness = 0 is normally distributed and not skewed. If there are more scores on the right-hand side of the distribution, then the distribution is negatively skewed. If there are more scores on the left-hand side of the distribution, then the distribution is positively skewed.

Slope   The change in Y divided by the change in X.

s_R^2 Between-row variance (reflects natural variation in the population and any variability due to your row variable).

s_{RC}^2 Row × Column variance (reflects natural variation in the population and any variability due to the interaction between your row and column variables).

Standard deviation   The average deviation of a score from the mean.

Statistic   A summary calculation on data collected from a sample.

s_w^2 Within-group variance (reflects natural variation in the population, and thus it is an estimate of σ^2). Within-group variance could also be called "error" (E) because it is variability that we cannot control but occurs naturally.

Symmetry   A distribution that is identically shaped on either side of the mean, or mirror images of one another.

Theoretical probability   The number of events classifiable as A divided by the total number of possible events.

Type I error   Error that occurs when you reject the null hypothesis when it is actually true.

Type II error   Error that occurs when you fail to reject the null hypothesis when it is actually false.

Variable   Any property that may take on different values at different times and may change with various conditions.

Variance   The squared average deviation of a score from the mean.

Y-intercept   The value of Y when X is equal to zero, which is where the line crosses the Y-axis.

References

Adams, D. C., & Anthony, C. D. (1996). Using randomization techniques to analyse behavioural data. *Animal Behavior, 51,* 733–738.

Craik, F. I. M., & Lockhart, R. S. (1972). Levels of processing: A framework for memory research. *Journal of Verbal Learning and Verbal Behavior, 11,* 671–684.

Cramer, H. (1946). *Mathematical methods of statistics.* Princeton, NJ: Princeton University Press.

Cranor, C. F. (1990). Some moral issues in risk assessment. *Ethics, 101,* 123–143.

Good, I. J. (1973). What are degrees of freedom? *The American Statistician, 27,* 227–228.

Good, P. (2006). *Resampling methods* (3rd ed.). Boston: Birkhauser.

Good, P. I. (2009). *Permutation, parametric, and bootstrap tests of hypotheses.* New York: Springer.

Harris, R. J. (1997). Significance tests have their place. *Psychological Science, 8,* 8–11.

Loftus, G. R. (1991). On the tyranny of hypothesis testing in the social sciences. *Contemporary Psychology, 36,* 102–104.

Loftus, G. R. (1993). A picture is worth a thousand *p* values: On the irrelevance of hypothesis testing in the microcomputer age. *Behavioral Research Methods, Instruments, and Computers, 25,* 250–256.

Loftus, G. R. (1996). Psychology will be a much better science when we change the way we analyze data. *Current Directions in Psychological Science, 5,* 161–171.

Nanna, M. J., & Sawilowsky, S. S. (1998). Analysis of Likert scale data in disability and medical rehabilitation research. *Psychological Methods, 3,* 55–67.

Smith, R. E., & Smoll, F. L. (1990). Self-esteem and children's reactions to youth sport coaching behaviors: A field study of self-enhancement processes. *Developmental Psychology, 26,* 987–993.

Index

SAGE Research Methods Online

The essential tool for researchers

Sign up now at
www.sagepub.com/srmo
for more information.

An expert research tool

- An **expertly designed taxonomy** with more than 1,400 unique terms for social and behavioral science research methods

- **Visual and hierarchical search tools** to help you discover material and link to related methods

- Easy-to-use navigation tools

- Content organized by complexity

- Tools for citing, printing, and downloading content with ease

- Regularly updated content and features

A wealth of essential content

- The most comprehensive picture of quantitative, qualitative, and mixed methods available today

- More than **100,000 pages of SAGE book and reference material** on research methods as well as editorially selected material from SAGE journals

- More than **600 books** available in their entirety online

Launching 2011!

SAGE research methods online